Springer Series On Environmental Management

Bruce N. Anderson
Robert W. Howarth
Lawrence R. Walker

Series Editors

Springer Series on Environmental Management
Volumes published since 1989

The Professional Practice of Environmental Management (1989)
R.S. Dorney and L. Dorney (eds.)

Chemical in the Aquatic Environment: Advanced Hazard Assessment (1989)
L. Landner (ed.)

Inorganic Contaminants of Surface Water: Research and Monitoring Priorities (1991)
J.W. Moore

Chernobyl: A Policy Response Study (1991)
B. Segerstahl (ed.)

Long-Term Consequences of Disasters: The Reconstruction of Friuli, Italy, in Its International Context, 1976–1988 (1991)
R. Geipel

Food Web Management: A Case Study of Lake Mendota (1992)
J.F. Kitchell (ed.)

Restoration and Recovery of an Industrial Region: Progress in Restoring the Smelter-Damaged Landscape near Sudbury, Canada (1995)
J.M. Gunn (ed.)

Limnological and Engineering Analysis of a Polluted Urban Lake: Prelude to Environmental Management of Onondaga Lake, New York (1996)
S.W. Effler (ed.)

Assessment and Management of Plant Invasions (1997)
J.O. Luken and J.W. Thieret (eds.)

Marine Debris: Sources, Impacts, and Solutions (1997)
J.M. Coe and D.B. Rogers (eds.)

Environmental Problem Solving: Psychosocial Barriers to Adaptive Change (1999)
A. Miller

Rural Planning from an Environmental Systems Perspective (1999)
F.B. Golley and J. Bellot (eds.)

Wildlife Study Design (2001)
M.L. Morrison, W.M. Block, M.D. Strickland, and W.L. Kendall

Selenium Assessment in Aquatic Ecosystems: A Guide for Hazard Evaluation and Water Quality Criteria (2002)
A.D. Lemly

Quantifying Environmental Impact Assessments Using Fuzzy Logic (2005)
R.B. Shephard

Changing Land Use Patterns in the Coastal Zone: Managing Environmental Quality in Rapidly Developing Regions (2006)
G.S. Kleppel, M.R. DeVoe, and M.V. Rawson (eds.)

The Longleaf Pine Ecosystem: Ecology, Silviculture, and Restoration (2006)
S. Jose, E.J. Tokela, and D.L. Miller (eds.)

Linking Restoration and Ecological Succession (2007)
L.R. Walker, J. Walker, R.J. Hobbs (eds.)

Lawrence R. Walker
Joe Walker
Richard J. Hobbs

Editors

Linking Restoration and Ecological Succession

Editors:

Lawrence R. Walker
Department of Biological Sciences
University of Nevada Las Vegas
Las Vegas, NV 89154
USA
walker@unlv.nevada.edu

Joe Walker
CSIRO Land and Water
PO Box 1666
Canberra ACT 2611
Australia
joe.walker@csiro.au

Richard J. Hobbs
School of Environmental Science
Murdoch University
Murdoch WA 6150
Australia
r.hobbs@murdoch.edu.au

Series Editors:

Bruce N. Anderson
Planreal Australasia
Keilor, Victoria 3036
Australia
bnanderson@compuserve.com

Robert W. Howarth
Program in Biogeochemistry
  and Environmental Change
Cornell University, Corson Hall
Ithaca, NY 14853
USA
rwh2@cornell.edu

Lawrence R. Walker
Department of Biological Sciences
University of Nevada Las Vegas
Las Vegas, NV 89154
USA
walker@unlv.nevada.edu

Cover Illustration: The photo shows several stages in the restoration of areas mined for phosphate on Christmas Island situated in the Indian Ocean.

1. The rainforest type in the background is complex mesophyll vine forest and the canopy height is about 35m.

2. The closed canopy vegetation in the middle distance is natural regrowth (i.e., a secondary succession) after about 25 years on areas cleared for phosphate mining—the canopy is dominated by 15 to 20m tall *Macaranga tanarius* (Euphorbiaceae), a common early successional tree on disturbed sites in the SE Asia/Pacific region.

3. The vegetation in the foreground was planted onto the mined area and is one year old. *Macaranga tanarius* (the large-leafed plant) is a major component of the rehabilitation and is used to mimic the early stages of natural succession. Photo by Paul Reddell.

Library of Congress Control Number: 2006927706

ISBN-10: 0-387-35302-X   e-ISBN-10: 0-387-35303-8
ISBN-13: 978-0387-35302-9   e-ISBN-13: 987-0387-35303-6

Preface

This book was conceived over dinner at the Cooloola Dunes near Brisbane, Australia, as we pondered how to reconcile 700,000 years of soil development with typical successional studies of <200 years and restoration concerns that normally cover <20 years. Restoration ecology is deeply rooted in ecological succession yet seems, as a fast-emerging discipline, to be largely unaware of the potential benefits a closer examination of succession can provide. These benefits address both how to restore ecosystem function and structure as quickly as possible and the longer-term consequences of current restoration activities. Successfully restored ecosystems can be more or less sustainable without constant care. This state is only achievable within a framework that recognizes, implicitly or explicitly, the temporal dynamics that constitute successional processes. While the current goals of restoration do not address change over thousands of years, certainly 2–200 year dynamics, the most common temporal scale for successional studies, are essential to consider. Restoration tactics will also differ depending on the age of the ecosystem being restored. Succession offers insights into processes of change in ecosystems of all ages, from very young, recently disturbed sites to very old systems such as the Cooloola Dunes.

Restoration ecology incorporates many areas of knowledge both within and outside traditional ecology. Succession often complements or reinforces these ties. Disturbance ecology is central to defining the physical limits for both succession and restoration. Landscape ecology, like restoration, operates within a spatial context and incorporates many ecosystems while succession offers more ecosystem-specific lessons. Studies of ecological assembly seek generalizations similar to succession and critical to the initiation of restoration. Invasion biology studies emerging ecosystems that both restoration and succession must address in a rapidly changing world. Studies of ecosystem health help define appropriate restoration goals but are rarely addressed in a successional context. Historical ecology provides proper land-use context for both restoration and succession. We argue that restoration within a successional framework will best utilize the lessons from each of these areas. Restoration, unlike successional studies, must cross disciplines and address societal needs, including politics, economics, human health issues, sustainability, and land-use planning.

Restoration also has much unfilled potential to elucidate fundamental unknowns within successional studies. When restoration is conducted within a scientific framework of replicated studies and peer-reviewed publication, it can clarify much about species change. Restoration is the acid test of our ability to understand not only how ecosystems are assembled and held together but also how they change over time. Proper documentation of both the failures and the successes of restoration activities will advance our understanding of many of the aforementioned subdisciplines of ecology, including succession, particularly within landscape gradients and novel, emergent ecosystems.

We assembled this book in order to examine and strengthen both the theoretical and practical ties between succession and restoration. We are not constrained by occasional differences in temporal or spatial scales between the two disciplines or the relative focus on natural versus human-managed ecosystems because restoration is fundamentally the management of succession. Restoration must ultimately succeed if disturbed landscapes are to be recovered and we argue that success will improve where successional principles are employed.

Lawrence was supported by a sabbatical from the University of Nevada Las Vegas, by Landcare Research in Lincoln, New Zealand, and by grants DEB-0080538 and DEB-0218039 from the National Science Foundation, as part of the Long Term Ecological Research Program in Puerto Rico.

We wish to thank the external reviewers whose comments made our jobs easier. These generous people include: Joe Antos, Peter Bellingham, Lisa Belyea, Ray Callaway, Vic Claassen, Viki Cramer, Tim Ellis, Valerie Eviner, Tadashi Fukami Ari Jumpponen, Werner Härdtle, David Mackenzie, Scott Meiners, Robin Pakeman, Gert Rosenthal, Simon Veitch, Evan Weiher, and Sue Yates, Truman Young, Joy Zedler. In addition, most chapter authors contributed reviews of one or more chapters.

Writing and editing books inevitably takes us away from many urgent family matters. The editors express their appreciation for the patience and support given by our wives Elizabeth, Janet, and Gillian, and the Hobbs' children Katie and Hamish during the long course of working on this book.

Lawrence R. Walker
Joe Walker
Richard J. Hobbs

Contents

Contributing Authors

Jan P. Bakker, Professor of Coastal Conservation Ecology, Community and Conservation Ecology Group, University of Groningen, PO Box 14, NL-9750 AA Haren, The Netherlands. j.p.bakker@rug.nl Jan studies coastal ecosystems with a focus on the restoration and management of marshes and grasslands.

Rudy van Diggelen, Assistant Professor of Plant Ecology, Community and Conservation Ecology Group, University of Groningen, PO Box 14, 9750 AA Haren, The Netherlands. r.van.diggelen@rug.nl Rudy studies community assembly and the functioning and restoration of degraded ecosystems.

Klauss Dierßen, Professor of Plant Ecology, Ecology Centre, University of Kiel, Olshausenstrasse 75, D-24118, Kiel, Germany. kdierssen@ecology.uni-kiel.de Klauss examines biodiversity, ecosystem functioning, and landscape planning.

Richard J. Hobbs, Professor of Environmental Science, School of Environmental Science, Murdoch University, Murdoch, WA 6150, Australia. r.hobbs@murdoch.edu.au Richard studies the management and restoration of altered ecosystems and landscapes.

Kai Jensen, Professor of Applied Plant Ecology, Population and Vegetation Ecology, Biocentre Klein Flottbek, University of Hamburg, Ohnhorststrasse 18, 22609 Hamburg, Germany. kai.jensen@botanik.uni-hamburg.de Kai's main interest is the vegetation of riparian and coastal wetlands and its relation to land-use as well as to abiotic site conditions and biotic interactions.

Anke Jentsch, Vegetation Ecologist, Centre for Environmental Research Liepzig, Department of Conservation Biology, UFZ, Permoserstr. 15, D-03418, Leipzig, Germany. anke.jentsch@ufz.de Anke's main research interests are in disturbance ecology, system dynamics, plant biodiversity, and ecosystem function.

Rob Marrs, Professor of Applied Plant Biology, Applied Vegetation Dynamics Laboratory, University of Liverpool, Liverpool L69 7ZB, United Kingdom. calluna@liverpool.ac.uk Rob's research interests focus on the manipulation of succession toward specified endpoints, specifically in heaths, moors, and bracken-dominated habitats.

Roger del Moral, Professor of Biology, Department of Biology, University of Washington, 1521 NE Pacific St., Seattle, Washington 98195-5325, USA. moral@u.washington.edu Roger studies the recovery of systems after catastrophic disturbances and practices wetland restoration.

Felix Müller, Systems Ecologist, Ecology Centre, University of Kiel, Olshausenstrasse 75, D-24118, Kiel, Germany. fmueller@ecology. uni-kiel.de Felix Mueller is a systems ecologist whose main research interests are in ecosystem theory, ecological indicators, and landscape analysis.

Duane Peltzer, Research Scientist, Landcare Research, PO Box 69, Lincoln 8152, New Zealand. peltzerd@landcareresearch.co.nz Duane studies interactions among plants and soil communities along environmental gradients and quantifies plant impacts on ecological and ecosystem processes.

Karel Prach, Professor of Botany, Faculty of Biological Sciences, University of České Budějovice, Branišovská 31, CZ-370 01 České Budějovice, Czech Republic AND Researcher, Institute of Botany, Academy of Sciences of the Czech Republic, Třeboň. prach@bf.jcu.cz Karel's main research interests concern vegetation succession, plant invasions, and ecology of river floodplains, with emphasis on restoration aspects.

Petr Pyšek, Researcher, Institute of Botany, Academy of Sciences of the Czech Republic, CZ-252 43, Průhonice, Czech Republic, AND Associate Professor, Department of Ecology, Faculty of Science, Charles University, Viničná 7, CZ-128 01 Praha 2, Czech Republic. pysek@ibot.cas.cz Petr's research interests include general aspects of biological invasions, ecology of invasive plant species, factors determining species diversity, and vegetation succession.

Paul Reddell, Chief Scientist, Ecobiotics Ltd., P.O. Box 148, Yungaburra, Queensland, Australia. paul.reddell@ecobiotics.com.au   Paul is a rainforest ecologist. He works for a company that searches for medical compounds in tropical rainforests. His research interests include rainforest dynamics and restoration of mined sites and tropical landscapes subjected to a wide range of disturbances.

Andreas Rinker, CEO, DygSyLand, Institute for Digital System Analysis and Landscape Diagnosis, Zum Dorfteich 6, 24975 Husby, Germany. rinker@digsyland.de Andreas' research interests focus on ecosystem modelling and landscape management.

Peter Schwartze, Geobotanist, Biologische Station Kreis Steinfurt e.V., Bahnhofstrasse 71, 49545 Tecklenburg, Germany. biologische.station. steinfurt@t-online.de   Peter's main research interests are on long-term development and management of grassland ecosystems.

Joachim Schrautzer, Assistant Professor, Ecology Centre, University of Kiel, Olshausenstrasse 75, D-24118, Kiel, Germany. jschrautzer@ecology.uni-kiel.de Joachim's main research interests focus on the management and restoration of fen and grassland ecosystems.

Vicky Temperton, Plant Ecologist, Phytosphere Institute, Juelich Research Centre, D-52425 Juelich, Germany. v.temperton@fz-juelich.de Vicky works on plant community assembly and its relation to restoration ecology and disturbance, as well as its influence on plant biodiversity and ecosystem function.

Joe Walker, Research Fellow, CSIRO Land and Water, PO Box 1666, Canberra, ACT 2611, Australia. joe.walker@csiro.au Joe is an eco-hydrologist whose main research interests include assessments of catchment and landscape condition, restoration of degraded agricultural lands, fire management, succession theory, and ecological indicators.

Lawrence R. Walker, Professor of Biology, Department of Biological Sciences, University of Nevada Las Vegas, 4505 Maryland Parkway, Las Vegas, Nevada 89154, USA. walker@unlv.nevada.edu Lawrence studies mechanisms and theories of plant succession and restoration ecology, and emphasizes cross-site comparisons.

David Wardle, Professor of Ecology, Department of Forest Vegetation Ecology, Swedish University of Agricultural Sciences, SE901-83 Umeå, Sweden AND Research Scientist, Landcare Research, PO Box 69, Lincoln 8152, New Zealand. david.wardle@svek.slu.se David studies links between aboveground and belowground communities, drivers of ecosystem functions, and soil ecology.

Forging a New Alliance Between Succession and Restoration

Lawrence R. Walker, Joe Walker, and Roger del Moral

Key Points

1. Succession and restoration are intrinsically linked because succession comprises species and substrate change over time and restoration is the purposeful manipulation of that change.
2. During succession both orderly and unpredictable patterns emerge but some general rules offer theoretical and practical insights for restoration activities. These insights are not often utilized due to inadequate communication and a misconception that because restoration is focused on shorter temporal scales and is more goal-oriented, then concepts from succession may not apply.
3. Restoration potentially offers succession practical insights into how communities assemble, but a dearth of scientific protocols in the conduct of restoration has hindered this linkage.

1.1 Introduction

How does succession take place, after all, and what are the adaptive cycles, if any, and the feedback systems, assembly rules and other inherent functional, evolutionary or simply dynamic mechanisms that make ecosystems develop and interact in one way or another? If we can sort these questions out—biome by biome—then we will unquestionably be better placed to predict how much time, energy and capital of all sorts will be required, or should be allocated, to ecological restoration and rehabilitation. (Aronson and van Andel 2006)

Human impacts on our planet are increasing exponentially, endangering our lifestyles and our survival. By one estimate, we would need another whole planet to provide humans in developing countries with the resource base currently exploited by the developed nations (Wackernagel and Rees 1996). One positive approach to ameliorate ecological impacts such as habitat loss and environmental degradation involves the rapidly developing field of restoration ecology. Within the past few decades, practitioners have formed an initial body of theory, an international society, and several professional journals that are helping them to organize the many examples of successful and unsuccessful efforts to restore damaged ecosystems. Several recent books have explored the theoretical basis of restoration ecology (Walker and del Moral 2003, Temperton *et al*. 2004, van Andel and Aronson 2006) and have better defined where it sits in the integrated world of ecology.

Restoration ecology is a multidisciplinary approach that implements in a practical way concepts drawn from a wide range of disciplines, including conservation biology, disturbance ecology, ecological succession, ecohydrology, invasion biology, island biogeography, and landscape ecology (Zedler 2005, Young *et al*. 2005, van Diggelen 2006). In addition, restoration ecology incorporates many other ecological themes, for example, biodiversity, habitat heterogeneity, resilience, and sustainability. Restoration often addresses political, economic, and sociological issues as well. As a new subdiscipline of ecology, restoration ecology has been driven primarily by the urgency to repair damaged landscapes. However, the success of restoration ecology in the practical realm will depend on the strength of the ecological and process-based underpinnings. One key link is with ecological succession, a central concept in ecology since Warming (1895) and Cowles (1899 and 1901) recognized that species change was related to the time since stabilization on dunes. Succession theory now encompasses a large set of concepts useful for explaining the mechanisms of plant community and ecosystem development (Glenn-Lewin *et al*. 1992, Walker and del Moral 2003). Restoration is fundamentally the manipulation of succession and frequently focuses on acceleration of species and substrate change to a desired endpoint (Luken 1990). While successful restoration intentionally repairs the processes driving succession, most other studies of vegetation change (e.g., global climate change, invasion biology, consequences of regional and watershed degradation, and gap dynamics; Davis *et al*. 2005) focus on the unintended factors that disrupt succession (Fig. 1.1). Yet, restoration often proceeds with more reliance on engineering, horticulture, and agronomy than on ecology (Young *et al*. 2005). Is this because succession does not have the answers, because restoration does not need succession, or because of a lack of communication or irreconcilable differences between the two disciplines?

This book contends that the overlap between restoration and succession has yet to be adequately explored and that restoration and ecological succession can and should forge a stronger alliance than exists today. Restoration will develop more coherently if it better integrates ideas generated from a century of studies of ecological succession. Additionally, restoration has great but under-utilized potential to help elucidate the fundamental processes controlling ecological succession by monitoring how key system drivers respond to treatments. In this book, a range of restoration types will be identified that differ in spatial scales, ecological drivers, and restoration goals. Data will be presented from restoration activities around the world that come from habitats representing gradients of precipitation, temperature, soil age, and the stability, fertility, and toxicity of substrates. By exploring such contrasting environments, we will seek generalizations that link successional theory to the practice of restoration.

A central question of this book is: "What is the minimum amount of biophysical and successional information needed to restore a specific landscape or area?" In addition, we can ask: "What target values or indicators offer the means to evaluate the relative success or progress of particular restoration strategies?"

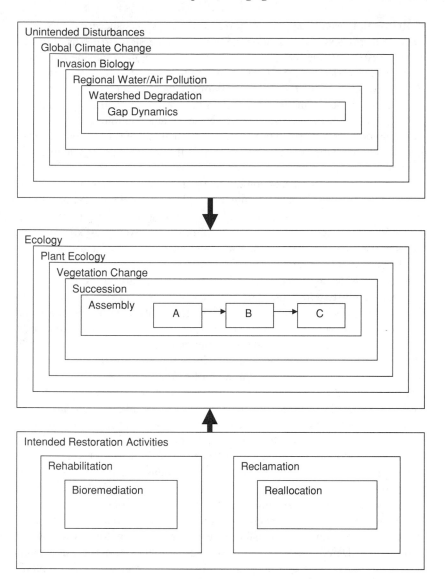

Figure 1.1 Plant succession (central box where community A proceeds to community C over time) can be impacted by unintended disruptions originating from disturbances that range in scale from global to local (top box) or by intended manipulations coming from restoration actions (bottom box). Restoration includes reclamation (any site amelioration), reallocation (alteration to a new function), rehabilitation (repair of ecosystem function), and bioremediation (reduction of site toxicity).

We will try to define what part of that information comes from successional theory and which key environmental drivers can be used to improve and measure restoration success. The urgency to repair damaged landscapes makes it critical to search for generalizations about the process of restoration. We suggest that examining restoration in the light of succession will aid in this search. Both of these topics are central to ecology—succession as the study of temporal dynamics and restoration as a critical application of ecological principles to an urgent societal need.

Restoration usually addresses shorter time scales than successional studies, but is, nevertheless, dependent on the broader successional patterns of change for its success (Palmer *et al.* 1997). Practicing restoration outside the framework of succession may be likened to building bridges without attention to the laws of physics. Patterns of successional development can offer reference systems for the assessment of restoration actions and critical insights into the roles of species dispersal, species interactions, plant–soil interactions, and soil development. In turn, restoration offers a practical test of successional theories in developing a stable, restored system within the constraints of socioeconomic demands.

Both restoration and succession can focus on structure and composition (e.g., the vertical distribution and accumulation of biomass and species) or function (e.g., ecosystem processes such as the flow of energy or cycling of materials). However, while succession is generally confined to a given ecosystem, restoration may address broader spatial scales and encompass adjacent ecosystems, catchments, and landscapes.

When considering ways to restore a habitat, landscape, ecosystem, or watershed, we must address questions about why, where, how, what, and when. The reasons why we want to restore are largely influenced by societal values and the economic imperative for sustainable resources and services, and are addressed by social economics and governmental policies (Costanza *et al.* 1992). Where to restore may be self-evident, but in many cases, the targeted area (e.g., a waterlogged area) is merely a symptom of a broader problem (e.g., extensive tree clearing and increased recharge into the water table) that needs attention at a larger spatial scale. How to successfully restore an area depends largely on the level of understanding of the main drivers of the overall system, a clear definition of endpoints, the level of degradation (perhaps a new system state), available technologies, and economic constraints (Walker and Reuter 1996). What is restored is determined by existing conditions, social attitudes, political and economic demands, and by ecological constraints. Biodiversity, stability, and ecosystem function are linked to the particular restoration goals such as achieving a particular species composition, ensuring the duration of a certain community type, and the provision of necessary or desired ecosystem goods and services. Determining when to restore is complex and the decision can be constrained by competing demands for funding or perceived threats to plants, animals, and humans. From an ecological perspective, many sequential actions over a long time-period are more likely to yield desired results than a single action. Such incremental change over time defines ecological succession. Finally, the evaluation of what constitutes successful restoration is most informative when placed in the broader ecological context of expectations based on knowledge about succession.

In the next sections, we first define our perceptions about succession and restoration and give a brief overview of each discipline. Then we discuss how succession and restoration differ in scale, subject matter, and underlying paradigms. Next, we explore in more detail how succession and restoration are similar, what each discipline has to offer the other, and how they are both limited by a common set of abiotic and biotic constraints. Finally, we introduce a focal set of questions and provide a summary of each of the following chapters.

1.2 Concepts

1.2.1 Definitions

Succession is the change in species composition and associated substrate changes over time. It is a dynamic process that is studied with descriptive, experimental, theoretical, and modeling approaches (McIntosh 1985). Formal, descriptive studies of succession began in the late 19th and early 20th centuries (Warming 1895, Cowles 1899 and 1901, Clements 1916), and were an extension of observations by natural historians, foresters, and agriculturalists during the previous several centuries (e.g., that ditches fill in with sediment and plant growth, that abandoned pastures gradually become forests, that stabilized dunes are colonized by plants). Experimental work began in the middle of the 20th century (Keever 1950) and continues to explore the mechanistic basis for species change [(reviewed as "neo-reductionism" in Walker and del Moral (2003)]. Theoretical studies of succession have had peaks of activity in the 1920s and 1930s (Ramensky 1924, Gleason 1926, Clements 1928), and again later in the century, emphasizing holism (Odum 1969), species life histories (Drury and Nisbet 1973, Huston and Smith 1987), reductionist models (Connell and Slatyer 1977, Pickett *et al.* 1987, Walker and Chapin 1987), and process-based computer models (Shugart and West 1980). Some current topics in succession include facilitation (Holmgren *et al.* 1997, Bruno *et al.* 2003, Walker *et al.* 2003), competition (Walker *et al.* 1989, Wilson 1999, Walker and del Moral 2003), herbivory (Davidson 1993, Fagan and Bishop 2000, Bishop *et al.* 2005), invasive species (Sheley and Krueger-Mangold 2003, Davis *et al.* 2005, Reinhart *et al.* 2005), priority effects (Samuels and Drake 1997, Corbin and D'Antonio 2004, Daehler and Goergen 2005), urban dynamics (Sukopp and Starfinger 1999, Robinson and Handel 2000, Sukopp 2004) and plant–soil interactions at both short-term (De Deyn *et al.* 2003, De Deyn *et al.* 2004, Bardgett *et al.* 2005) and long-term scales (Walker *et al.* 2001, Hedin *et al.* 2003, Wardle *et al.* 2004). Many of these topics are shared by other subdisciplines within ecology but succession is specifically concerned with their influence on temporal dynamics. There is some agreement on the basic principles that drive succession but still no overarching paradigm for the myriad possible outcomes of succession (McIntosh 1999, Walker and del Moral 2003).

Restoration, in a broad sense, is the manipulation of a disturbed habitat or landscape to a desired condition. It is therefore more focused on specific outcomes than studies of succession, which attempt to understand the nature of vegetation change. Restoration has been part of agricultural and forestry activities and other human impacts on ecosystems for a long time, as with shifting agriculture or efforts to replenish eroded soils. Restoration ecology attempts to bring some ecological principles into restoration actions and focuses almost entirely on habitats impacted by or relevant to human activities. Restoration ecology is a more practical management science than the study of succession and is more integrated with socioeconomic and political realities. Restoration ecology is also a younger scientific discipline than succession, with conceptual origins in the 1940s (e.g., Leopold 1949), but formalization as a field of study only in the 1980s (Bradshaw and Chadwick 1980, Cairns 1980, Jordan *et al.* 1987). While still largely descriptive, each restoration action is, in practice, an experiment. Integration of experimentation and restoration activities with the intent to seek generalization is increasingly common (Dobson *et al.* 1997, Gilbert and Anderson 1998, Zedler and Callaway 2003). Formal development of the theoretical basis of restoration ecology is, however, still in its early stages (Cairns and Heckman 1996, Hobbs and Norton 1996, Young *et al.* 2005).

In this book, we will use very broad definitions of succession and restoration. In this way, succession encompasses severely damaged or new substrates (primary) and more intact ones (secondary). Succession also addresses many possible trajectories (e.g., retrogressive, direct regeneration, divergent, convergent) and types of organismal change (e.g., of animals, plants, or microbes). We are also not limited by the disturbance type that initiates succession, although the bias will be toward disturbances of most relevance to humans and restoration efforts. Restoration will be used in the broadest sense as well (Fig. 1.1; Aronson *et al.* 1993), incorporating reclamation (any site amelioration), reallocation (alteration to a new function), rehabilitation (repair of ecosystem function), and bioremediation (reduction of site toxicity). Our use of restoration *sensu lato* does not encompass the full recovery of an ecosystem to its pre-disturbance structure and function (restoration *sensu stricto*), as we regard that goal as generally unrealistic.

1.2.2 Differences

Succession and restoration differ in scale, subject matter, and underlying paradigms. Succession most commonly addresses time intervals between 10 and 200 years, encompassing the life times of most perennial vascular plants. Restoration typically focuses on periods between 1 and 20 years, or the duration of human involvement in most projects. Both can, of course, address a wider range of temporal scales, particularly successional trajectories that can extend to thousands of years (Walker *et al.* 1981, Crews *et al.* 1995, Wardle *et al.* 2004). The use of chronosequences, or space for time substitutions (Pickett 1989), is essential for longer time scales and is well established in successional studies but is not commonly used for restoration planning (Walker *et al.* 2001, Hobbs 2005). Spatial scales can differ also, with succession often focusing at smaller scales.

The subjects that the field of succession most often addresses are tightly linked to either specific disturbances (often natural ones) or the disturbance regime (the composite of all disturbances in a region). The link to humans is not a prerequisite, although more successional studies address changes following agriculture or logging (Glenn-Lewin *et al.* 1992) than natural processes such as glacial melt or volcanic eruptions (Walker and del Moral 2003). In contrast, most studies of restoration tackle only those disturbances most relevant to humans. Succession focuses on changes within one successional sequence (sere), and remains within one ecosystem. Restoration, on the other hand, often addresses adjoining ecosystems, such as those within a watershed, urban area, or landscape (Holl *et al.* 2003). In this way, restoration can encompass multiple seres such as those that follow from agriculture and logging within the same landscape (van Diggelen 2006).

It is evident throughout this account that many of the basic paradigms of succession and restoration differ substantially. Succession, with its roots in natural history and observations about habitat changes over time, has developed a conceptual framework entrenched in scientific methodology and motivated to understand mechanisms of species change. This is a classic supply paradigm, with a proliferation of information that may someday be useful. Restoration has developed closer links to the practical concerns of managers and is more action-oriented with motivation to achieve particular results. This is a classic demand paradigm, with the practical issues of day-to-day management demanding sustainable practices based on environmental knowledge. Restoration ecology has begun to develop a stronger conceptual framework but the application of successional studies to practical management issues is still inadequate.

1.2.3 Similarities and Linkages

Despite the different origins and approaches noted above, succession and restoration share many traits that make stronger linkages an achievable proposition (Table 1.1). Both are concerned with responses to disturbance (especially human-initiated ones). They both deal with a subset of the landscape and are dependent on knowledge about ecosystem function, community structure and dynamics, and species attributes in order to proceed. In addition, both are concerned with the modeling and prediction of the sequence of discrete events called successional trajectories. We next explore these linkages in terms of what each discipline offers the other.

1.2.3.1 Succession to Restoration

Because of a century of study in many areas of the globe, the discipline of succession can offer substantial contributions to the discipline of restoration. Succession offers both a long-term perspective and short-term predictions on species dynamics and provides a reference system for restoration that can suggest likely outcomes following management actions (Aronson and van Andel 2006). Methods developed within or used in studies of succession that are or can be incorporated into restoration include, for example, functional plant groups (vital attributes), species filters, ecosystem assembly, state and transition models, fuzzy set theory, Markov processes, and biogeochemical modeling. Succession contributes to an understanding that multiple trajectories are

Table 1.1 Topics that link succession and restoration. Studies of succession and restoration share much overlap in subject matter. Succession offers restoration insights into: responses to different disturbance regimes; how responses to various ecosystem functions reflect changes in community structure and dynamics as measured by species attributes; generalizations about possible trajectories; and models that predict possible outcomes of succession. Restoration offers succession practical data on amelioration of infertility and other abiotic constraints as well as input about species interactions in particular circumstances and the sustainability of various successional communities.

Topic	Shared Information
Disturbance	Loss of biological legacy (severity), disturbance regime
Ecosystem Function	Energy flow, carbon accumulation and storage, nutrient dynamics, soil properties, water cycle
Community Structure and Composition	Biomass, vertical distribution of leaf area, leaf area index, species richness, species evenness, species density, spatial aggregation
Community Dynamics	Facilitation, inhibition, dispersal (priority effects and entrapment), sustainability
Species Attributes	Life history characteristics (pollination syndrome, germination, establishment, growth, longevity)
Trajectories	Rates and targets
Models	Generalizations about processes

possible so restoration goals must remain flexible and open to change. Succession theory also suggests that reconstruction of dynamic ecosystems must incorporate responses to changes from within the system (typically from species interactions) and from disturbances from outside (typically from modifications to abiotic variables but also from biotic invasions). These kinds of changes, some predictable, others less so, mean that restoration must follow an adaptive management style (Zedler and Callaway 2003). Insights from succession can elucidate various ecosystem functions for restoration, including local hydrology, soil development, energy flow, nutrient dynamics, and carbon accumulation and storage (Table 1.1). At the plant community level, structural insights from succession include information on biomass, leaf area distribution, leaf area indices, species richness, species evenness, species density, and spatial aggregation that can help restoration ecologists. Community dynamics, a common topic for successional studies, helps explain the type (plant/soil, plant/plant, and plant/animal) and mode (facilitation, competitive inhibition) of species interactions as well as insights into dispersal (priority effects, entrapment). Successional studies also offer information about various life history characteristics (pollination, germination, establishment, growth, longevity) of key species as well as many other plant traits. Modeling species change can help generalize lessons about restoration from site-specific studies to reach more broadly applicable conclusions. Many of these concepts can help the people working on restoration projects to organize data collection, determine immediate restoration activities, estimate rates of change, and plan the long-term search for appropriate development of generalizations and understanding of mechanisms.

1.2.3.2 Restoration to Succession

Restoration studies potentially can provide a wealth of information to improve our understanding of succession. Practical tests of successional theory are obvious outcomes if the restoration activities use normal scientific protocols such as inclusion of a control, quantitative data collection, planned treatment comparisons, statistical analysis, and peer-reviewed publication of results. Restoration can also provide insights into both historical and biological links among landscape components as well as potential details on impacts to the water cycle or on substrate changes. Efforts to address soil toxicity and infertility, as well as efforts to promote propagule dispersal to the site, are activities that can contribute to understanding the physiology and life histories of key species. Restoration activities can also help inform us about community structure (species richness, evenness, density, spatial aggregation) and community sustainability. Species performance in restoration offers insights into life history characteristics, perhaps generating more information than strictly successional studies. Finally, restoration has much to teach succession by asking practical questions about trajectories and targets. For example, can certain successional stages be skipped in order to jump-start succession or hasten the establishment of a desired community?

1.3 Common Constraints

Succession and restoration are limited by a similar set of abiotic and biotic constraints that include plant dispersal, germination, and growth, as well as species turnover and ecosystem resilience. Overcoming these constraints is a major part of the task of any successful restoration. Successional studies can explain how such constraints are naturally overcome, both for short-term restoration tactics to establish vegetative cover and for long-term efforts to restore ecosystem resilience relative to the prevailing disturbance regime (Walker and del Moral 2003). Although site-specific solutions are the most dependable resolution to each set of interacting constraints, we offer a few generalizations below and some examples that illustrate how succession can assist restoration efforts in overcoming the constraints. Later chapters will expand on many of these topics.

1.3.1 Abiotic Factors and Their Amelioration

Generalizations about colonization, plant growth, and succession are highly dependent on climate and the nutritional and physical properties of the substrate. Vascular plant establishment and growth are frequently restricted by water availability (Cody 2000). Restoration in dry, cold conditions often involves addition of mulches to conserve water and promote mineralization such as on gold mine spoils in central Alaska (Densmore 1994). Restoration in dry, hot conditions can sometimes be improved by decompaction of soils to increase permeability, as on a chronosequence of abandoned roads in the Mojave Desert (Bolling and Walker 2000). Poor establishment and slow growth under either temperature extreme can be aided by microclimate amelioration using other vegetation as nurse plants to provide shade or windbreaks and surface contouring. However, the positive effect of a vegetative cover decreases as the environment becomes more favorable (Callaway and Walker 1997, Holmgren *et al.* 1997). Therefore, solutions appropriate to early stages of succession may not work in later, more fertile stages.

Substrate quality, measured by age, stability, fertility, and toxicity, can affect restoration success. Very old soils are often low in phosphorus (Walker and Syers 1976, Crews *et al.* 1995, Wardle *et al.* 2004). Poor nutritional status due to weathering and leaching (Gunn and Richardson 1979), structural decline, crusting, and the acidification of the topsoil are also characteristics of old soils (Russell and Isbell 1986). Restoration on older soils, such as are found in many parts of Australia, can be a very difficult proposition because of the long-term accumulation of salt from atmospheric accession and its subsequent mobilization (Williams *et al.* 2001).

Substrate stability depends on slope, rainfall amount and intensity, soil texture and erodability, vegetative cover, and grazing type, frequency, and intensity. Recurring disturbances will also impact stability and therefore restoration efforts. On landslides in Puerto Rico, restoration efforts include a variety of physical and biotic interventions from mulches to planting to recontouring the whole hillside, each adjusted to the disturbance regime of a particular landslide. Despite all these restoration efforts, the dense growth of native vine-like ferns that invade via clonal growth from the edges appear to be the best stabilizers (Walker 2005). The drawback with the ferns is that they can delay succession for several decades (Walker *et al.* 1996). Dunes provide another example of recurring disturbances where stabilization is critical before succession can proceed. Reduction of human impacts, planting rather than sowing grasses or fast-growing trees, and artificial stabilization are all possible methods of accelerating dune succession (Nordstrom *et al.* 2000).

Grazing is a major cause of soil erosion in such places as Iceland and China. In Iceland, 1000 years of heavy grazing have left little protection from wind, rain, and ice heaving. Native ground cover and fences to exclude grazers allow *Betula* forests to develop within 50 years (Aradottir and Eysteinsson 2004) but pressure from sheep farmers for open grazing land, especially in the vulnerable uplands, keeps most of Iceland deforested. The Loess Plateau in central China has experienced even longer agricultural activities than Iceland (>2000 years). Cultivation and the highly erodable, aeolian soils have led to severe erosion problems and huge sediment loads in the Yellow River (McVicar *et al.* 2002). The most actively eroded areas are being replaced with perennial vegetation. Some 150,000 km^2 of eroded land has been controlled by various conservation measures (Rui *et al.* 2002).

Substrate fertility is a very common constraint for restoration, especially where little or no topsoil remains. Loss of organic soil reduces nitrogen levels critical for revegetation (Classen and Hogan 2002). Organic layers can be eroded or leached where rainfall is sufficient such as on Himalayan landslides (Pandey and Singh 1985) or hardpans may develop at the surface where evaporation exceeds precipitation in arid lands (Zougmore *et al.* 2003). To restore such soils, a balance is needed between sufficient fertilization that promotes succession and excessive fertilization that favors strong competitors that reduce biological diversity and inhibit further succession (Prach 1994, Marquez and Allen 1996, Walker and del Moral 2003). For example, fertilization of nonnative grasses delayed recolonization of native tundra species on the Alaskan pipeline corridor by several decades (Densmore 1992). In such cases, restoration goals were limited to simply providing vegetative cover and failed to address species interactions and successional dynamics.

Another common constraint is toxic surfaces such as found on landfills and mine tailings. Under such conditions, restoration goals are usually relaxed and any cover is considered a success. Approaches include sealing the surface, topsoil and mycorrhizal additions, conversion to wetlands, sowing with grasses that are tolerant of the toxins, and planting trees (Bradshaw 1952, Wali 1992, Cooke 1999). Bioremediation, or the direct amelioration of toxic conditions with plants and microbes, is a growing field, but one that has not yet been incorporated into successional frameworks (Walker and del Moral 2003).

1.3.2 Establishment

The first stage of succession involves successful dispersal of plant reproductive units to the site of interest, while restoration involves deliberate sowing of those same plants or appropriate plant introductions, but usually after some site preparation. Preliminary assessment of the viability and quantity of natural seed rain can determine if introductions are needed. Where dispersal is determined to be a limiting factor, perches can encourage introduction of relatively heavy, bird-dispersed seeds, as on landslides in Puerto Rico (Shiels and Walker 2003). In other cases, assessment of pollination and seed production at the site or in the vicinity will also be required. For example, restoration of the perennial rosette *Argyroxiphium sandwicense* in Hawaii was improved by studies that determined that it was pollinator limited and largely self-incompatible (Powell 1992). These discoveries led to artificial pollination to increase seed set and clustered out-plantings to improve cross-pollination. Existing seed banks are another variable to evaluate, as they do not always reflect the existing or pre-disturbance vegetation. For example, in native Hawaiian forests most of the seed bank is composed of alien plants (Drake 1998). If no viable seeds exist in the soil or there is no seed rain, desirable species can be sown or transplanted to accelerate both restoration and succession (Pärtel *et al.* 1998). Germination and early survival are the final steps to establish initial populations. Germination requirements vary enormously among plants with the most known about species of agricultural interest. Conditions for seedling survival are more generic: protection from herbivory and adequate warmth, nutrients, and water. Vegetative reproduction can aid restoration, especially in primary succession, and bypass many of the constraints noted above (del Moral and Jones 2002). Successional contributions, aside from direct experience with particular species, will largely be to provide an overall demographic context for restoration efforts.

1.3.3 Growth and Species Interactions

Once plants have survived their first growing season at a site, new factors become the focus of both succession and restoration. Adequate growth will be assessed according to the goals of the restoration project. In most cases, rapid growth is desired. Belowground growth helps promote substrate stabilization and deters desiccation of seedlings while rapid growth aboveground can help reduce surface erosion and deter losses from ground-dwelling herbivores. Successional studies help identify bottlenecks to successful growth and potential effects of species interactions. Many studies of succession have focused on the relative balance between competition and facilitation and have direct relevance to restoration. One general lesson suggests that facilitation will be more important for species change (and restoration) in severe habitats while competition dominates more fertile, mesic, and stable habitats (Callaway and Walker 1997). However, many species interactions embody the whole suite of competitive to facilitative effects on each other (Bronstein 1994) and the relative balance of these can shift during the life span of each species (Bruno 2000, Walker *et al.* 2003), particularly as relative growth rates or sizes change (Callaway and Walker 1997). For example, the shrub *Mimosa luisana* initially facilitates establishment of the cactus *Neobuxbaumia tetetzo* but eventually the cactus inhibits the shrub's growth and reproduction (Flores-Martinez *et al.* 1994). In contrast, *Alnus sinuata* shrubs in Alaska initially inhibited germination of *Picea sitchensis* trees but the nitrogen added by *A. sinuata* facilitated later growth of *P. sitchensis* (Chapin *et al.* 1994). Species interactions, whether facilitative or inhibitory, are often site- and species-specific. Nevertheless, restoration activities can improve recruitment and survival rates when they incorporate the latest relevant information about species interactions.

1.3.4 Ecosystem Resilience

The ultimate goal of a restoration project is to have a self-sustaining community that includes natural species turnover and resilient responses to the local disturbance regime (Table 1.1). Both succession (Glenn-Lewin *et al.* 1992, Walker and del Moral 2003) and restoration (Temperton *et al.* 2004, van Andel and Aronson 2006) are moving away from the notion of a static climax community toward a more dynamic view of communities. In this dominant view, communities actively change in response to such internal processes as species change (driven largely by species longevity, competition, facilitation, or invasions) and external drivers including historical climate change (McGlone 1996) and disturbances that alter structure by damaging or removing biomass. Few restoration projects have recreated such ideal conditions. Indeed, those that come closest are probably those with the least human intervention. Nonetheless, many principles gleaned from studies of succession can provide a good foundation for restoration programs, including the restoration of novel ecosystems dominated by new combinations of native and nonnative species (Aronson and van Andel 2006) and the possibility of developing natural ecosystem mimicry in agricultural landscapes (Leroy *et al.* 1999).

1.4 Book Outline

Each chapter in this book will focus on a set of central questions in order to develop both new theoretical advances in the field of ecosystem restoration and practical tools to improve ecosystem management. These questions are:

1. What site and landscape factors are likely to determine and/or limit restoration?
2. What do observations from the study of succession offer for improved restoration practice?
3. How can restoration practices be improved across many sites and landscapes with the application of these successional concepts?
4. How can restoration practices inform our understanding of succession?

By addressing these questions for a wide variety of ecosystems we hope to accomplish the following goals:

1. Provide the latest understanding of linkages between successional theory and restoration practice.
2. Increase potential restoration effectiveness by providing instructive models from natural recovery processes.
3. Consider applications from local to landscape scales.
4. For the first time, consider landscape ages as key drivers linking succession and restoration.
5. Link the emphasis on general plant traits and results (e.g., ecosystem function) to succession and restoration.
6. Examine restoration and management of ecohydrological issues and how they are linked to succession.
7. Integrate the crucial role of soil biota as a means to manipulate trajectories with other aspects of system manipulation.

Chapter 2 (R. del Moral, L. Walker, J. Bakker) examines the lessons and insights gained from practical experiences in succession that can improve restoration results. The focus is on processes such as dispersal, germination, competition, and herbivory that can be easily manipulated and the bottlenecks that must be overcome in order to restore an optimal balance between ecosystem structure and function.

Chapter 3 (D. Wardle, D. Peltzer) examines the restoration of soil ecosystems within the context of fertility and soil biota. This chapter addresses the role of sequential changes that occur during the development of soils following disturbance and how to best manipulate and use soil biota as engineers of restoration success.

Chapter 4 (J. Walker, P. Reddell) asserts that succession and hence restoration endpoints on old landscapes differ from succession and restoration actions on young landscapes. This chapter highlights retrogressive succession and addresses how temporal and spatial scales impact the linkages between succession and practical restoration efforts. Old landscapes are contrasted with more recent surfaces to bridge many temporal scales. A tropical forest in northern Australia and salinized landscapes in semiarid Australia are used as examples.

Chapter 5 (J. Schrautzer, A. Rinker, K. Jensen, F. Müller, P. Schwartze, K. Dierßen) addresses the utility of broad ecosystem-based management and the contributions of successional concepts and catchment scale dynamics to restoration of European fens. Modeled values for key ecosystem variables are used to contrast retrogressive succession following increased disturbance intensity and progressive succession following abandonment of former fens.

Chapter 6 (K. Prach, R. Marrs, P. Pyšek, R. van Diggelen) explores the degree to which we can manipulate succession. When is it best to let succession proceed without intervention? When is it best to arrest succession? What must be done to integrate the reality of invasive species into restoration plans?

Chapter 7 (R. Hobbs, A. Jentsch, V. Temperton) explores the linkages between restoration and succession using the concepts of species assembly and disturbance. Do assembly rules and self-organizational principles help restoration planning and increase restoration effectiveness?

Chapter 8 (R. Hobbs, L. Walker, J. Walker) integrates the concepts covered in this book, attempts to answer our core set of questions, and explores how succession can assist restoration planning, our understanding of species trajectories, and temporal changes in ecosystem functions.

Acknowledgments: Comments by Peter Bellingham, Viki Cramer, Karel Prach, Vicky Temperton, and Sue Yates greatly improved this chapter. Lawrence Walker acknowledges sabbatical support from the University of Nevada Las Vegas and Landcare Research, New Zealand.

References

Aradottir, A. L., and Eysteinsson, T. 2004. Restoration of birch woodlands in Iceland. In: *Restoration of Boreal and Temperate Forests.* J. Stanturf and P. Madsen (eds.). Boca Raton: CRC/Lewis, pp. 195–209.

Aronson, J., and van Andel, J. 2006. Challenges for ecological theory. In: *Restoration Ecology.* J. van Andel and J. Aronson (eds.). Oxford, U.K.: Blackwell, pp. 223–233.

Aronson, J., Floret, C., LeFloc'h, E, Ovalle, C., and Pontanier, R. 1993. Restoration and rehabilitation of degraded ecosystems in arid and semiarid regions. I. A view from the South. *Restoration Ecology* 1:8–17.

Bardgett, R. D., Bowman, W. D., Kaufman, R., and Schmidt, S. K. 2005. A temporal approach to linking aboveground and belowground ecology. *Trends in Ecology and Evolution* 20:634–641.

Bishop, J. G., Fagan, W. F., Schade, J. D., and Crisafulli, C. M. 2005. Causes and consequences of herbivory on prairie lupine (*Lupinus lepidus*) in early primary succession. In: *Ecological Responses to the 1980 Eruption of Mount St. Helens.* V. H. Dale, F. J. Swanson, and C. M. Crisafulli (eds.). New York: Springer, pp. 151–161.

Bolling, J. D., and Walker, L. R. 2000. Plant and soil recovery along a series of abandoned desert roads. *Journal of Arid Environments* 46:1–24.

Bradshaw, A. D. 1952. Populations of *Agrostis tenuis* resistant to lead and zinc poisoning. *Nature* 169:1098.

Bradshaw, A. D., and Chadwick, M. J. 1980. *The Restoration of Land: The Ecology and Reclamation of Derelict and Degraded Land*. Oxford, U.K.: Blackwell.

Bronstein, J. L. 1994. Conditional outcomes in mutualistic interactions. *Trends in Ecology and Evolution* 9:214–217.

Bruno, J. F. 2000. Facilitation of cobble beach plant communities through habitat modification by *Spartina alterniflora*. *Ecology* 81:1179–1192.

Bruno, J. F., Stachowicz, J. J., and Bertness, M. D. 2003. Inclusion of facilitation into ecological theory. *Trends in Ecology and Evolution* 18:119–125.

Cairns, J. (ed.). 1980. *The Recovery Process in Damaged Ecosystems*. Ann Arbor, MI: Ann Arbor Science Publishers.

Cairns, J., and Heckman, J. R. 1996. Restoration ecology: The state of an emerging field. *Annual Review of Energy and Environment* 21:167–189.

Callaway, R. M., and Walker, L. R. 1997. Competition and facilitation: A synthetic approach to interactions in plant communities. *Ecology* 78:1958–1965.

Chapin, F. S., III., Walker, L. R., Fastie, C. L., and Sharman, L. C. 1994. Mechanisms of primary succession following deglaciation at Glacier Bay, Alaska. *Ecological Monographs* 64:149–175.

Claasen, V. P., and Hogan, M. P. 2002. Soil nitrogen pools associated with revegetation of disturbed sites in the Lake Tahoe area. *Restoration Ecology* 10:195–203.

Clements, F. E. 1916. *Plant Succession: An Analysis of the Development of Vegetation*. Washington, D.C.: Carnegie Institution of Washington Publication.

Clements, F. E. 1928. *Plant Succession and Indicators*. New York: H.W. Wilson.

Cody, M. 2000. Slow-motion population dynamics in Mojave Desert perennial plants. *Journal of Vegetation Science* 11:351–358.

Connell, J. H., and Slatyer, R. O. 1977. Mechanisms of succession in natural communities and their roles in community stability and organization. *The American Naturalist* 111:1119–1144.

Cooke, J. A. 1999. Mining. In: *Ecosystems of Disturbed Ground*. L. Walker (ed.). Amsterdam, The Netherlands: Elsevier, pp. 365–384.

Corbin, J. D., and D'Antonio, C. M. 2004. Competition between native perennial and exotic annual grasses: Implications for an historical invasion. *Ecology* 85:1273–1283.

Costanza, R., Norton, B. G., and Haskell, B. D. 1992. *Ecosystem Health: New Goals for Environmental Management*. Washington, D.C.: Island Press.

Cowles, H. C. 1899. The ecological relations of the vegetation on the sand dunes of Lake Michigan. 1. Geographical relations of the dune flora. *Botanical Gazette* 27:95–117, 167–202, 281–308, 361–391.

Cowles, H. C. 1901. The physiographic ecology of Chicago and vicinity: A study of the origin, development, and classification of plant societies. *Botanical Gazette* 31:73–108, 145–182.

Crews, T., Kitayama, K, Fownes, J., Riley, R. Herbert, D., Mueller-Dombois, D., and Vitousek, P. 1995. Changes in soil phosphorus fractions and ecosystem dynamics across a long chronosequence in Hawaii. *Ecology* 76:1407–1424.

Daehler, C. C., and Goergen, E. M. 2005. Experimental restoration of an indigenous Hawaiian grassland after invasion by Buffel grass (*Cenchrus ciliaris*). *Restoration Ecology* 13:380–389.

Davidson, D. W. 1993. The effects of herbivory and granivory on terrestrial plant succession. *Oikos* 68:23–35.

Davis, M. A., Pergl, J., Truscott, A.-M., Kollmann, J., Bakker, J. P., Domenech, R., Prach, K., Prieur-Richard, A.-H., Veeneklaas, R. M., Pyšek, P., del Moral, R., Hobbs,

R. J., Collins, S. L., Pickett, S. T. A., and Reich, P. B. 2005. Vegetation change: A reunifying concept in plant ecology. *Perspectives in Plant Ecology, Evolution and Systematics* 7:69–76.

De Deyn, G. B., Raaijmakers, C. E., and Van der Putten, W. H. 2004. Plant community development is affected by nutrients and soil biota. *Journal of Ecology* 92:824–834.

De Deyn, G. B., Raaijmakers, C. E., Zoomer, H. R., Ber, M. P., de Ruiter, P. C., Verhoef, H. A., Bezemer, T. M., and van der Putten, W. H. 2003. Soil invertebrate fauna enhances grassland succession and diversity. *Nature* 422:711–713.

del Moral, R., and Jones, C. C. 2002. Early spatial development of vegetation on pumice at Mount St. Helens. *Plant Ecology* 161:9–22.

Densmore, R. V. 1992. Succession on an Alaskan tundra disturbance with and without assisted revegetation with grass. *Arctic and Alpine Research* 24:238–243.

Densmore, R. V. 1994. Succession on regraded placer mine spoil in Alaska, USA, in relation to initial site characteristics. *Arctic and Alpine Research* 26:354–363.

Dobson, A. P., Bradshaw, A. D., and Baker, A. J. M. 1997. Hopes for the future: Restoration ecology and conservation biology. *Science* 277:515–522.

Drake, D. R. 1998. Relationships among the seed rain, seed bank and vegetation of a Hawaiian forest. *Journal of Vegetation Science* 9:103–112.

Drury, W. H., and Nisbet, I. C. T. 1973. Succession. *Journal of the Arnold Arboretum* 54:1147–1164.

Fagan, W. F., and Bishop, J. G. 2000. Trophic interactions during primary succession: Herbivores slow a plant reinvasion at Mount St. Helens. *The American Naturalist* 155:238–251.

Flores-Martinez, A., Ezcurra, E., and Sanchez-Colon, S. 1994. Effect of *Neobuxbaumia tetetzo* on growth and fecundity of its nurse plant *Mimosa luisana*. *Journal of Ecology* 82:325–330.

Gilbert, O. L., and Anderson, P. 1998. *Habitat Creation and Repair*. Oxford: Oxford University Press.

Gleason, H. A. 1926. The individualistic concept of the plant association. *Bulletin of the Torrey Botanical Club* 53:7–26.

Glenn-Lewin, D. C., Peet, R. K., and Veblen, T. T. (eds.). 1992. *Plant Succession: Theory and Prediction*. London: Chapman and Hall.

Gunn, R. H., and Richardson, P. D. 1979. The nature and possible origins of soluble salts in deeply weathered landscapes in eastern Australia. *Australian Journal of Soil Research* 17: 197–215.

Hedin, L. O., Vitousek, P. M., and Matson, P. A. 2003. Nutrient losses over four million years of tropical forest development. *Ecology* 84:2231–2255.

Hobbs, R. J. 2005. The future of restoration ecology: Challenges and opportunities. *Restoration Ecology* 13:239–241.

Hobbs, R. J., and Norton, D. A. 1996. Towards a conceptual framework for restoration ecology. *Restoration Ecology* 4:93–110.

Holl, K. D., Crone, E. E., and Schultz, C. B. 2003. Landscape restoration: Moving from generalities to methodologies. *BioScience* 53:491–502.

Holmgren, M., Scheffer, M., and Huston, M. A. 1997. The interplay of facilitation and competition in plant communities. *Ecology* 78:1966–1975.

Huston, M., and Smith, T. 1987. Plant succession: Life history and competition. *The American Naturalist* 130:168–198.

Jordan, W. R., Gilpin, M. E., and Aber, J. D. 1987. *Restoration Ecology: A SyntheticApproach to Ecological Research*. Cambridge: Cambridge University Press.

Keever, C. 1950. Causes of succession on old fields of the piedmont, North Carolina. *Ecological Monographs* 20:230–250.

Leopold, A. 1949. *A Sand County Almanac*. Oxford: Oxford University Press.

Leroy, E. C., Hobbs, R. J., O'Connor, M. H., and Pate, J. S. 1999. Agriculture as a mimic of natural systems. *Agroforestry Systems* 45(Special Issue):1–446.

Luken, J. O. 1990. *Directing Ecological Succession.* London: Chapman and Hall.

Marquez, V. J., and Allen, E. B. 1996. Ineffectiveness of two annual legumes as nurse plants for establishment of *Artemisia californica* in coastal sage scrub. *Restoration Ecology* 4:42–50.

McGlone, M. 1996. When history matters: Scale, time climate and tree diversity. *Global Ecology and Biogeography Letters* 5:309–314.

McIntosh, R. P. 1985. *The Background of Ecology.* Cambridge: Cambridge University Press.

McIntosh, R. P. 1999. The succession of succession: A lexical chronology. *Bulletin of the Ecological Society of America* 80:256–265.

McVicar, T. R., Rui, L., Walker, J., Fitzpatrick, R. W., and Changming, L. (eds.). 2002. *Regional Water and Soil Assessment for Managing Sustainable Agriculture in China and Australia.* Canberra: Australian Centre for International Agricultural Research.

Nordstrom, K. F., Lampe, R., and Vandemark, L. M. 2000. Reestablishing naturally functioning dunes on developed coasts. *Environmental Management* 25:37–51.

Odum, E. P. 1969. The strategy of ecosystem development. *Science* 164:262–270.

Palmer, M. A., Ambrose, R. F., and Poff, N. L. 1997. Ecological theory and community restoration ecology. *Restoration Ecology* 5:291–300.

Pandey, A. N., and Singh, J. S. 1985. Mechanism of ecosystem recovery: A case study from Kumaun Himalaya. *Recreation and Revegetation Research* 3:271–292.

Pärtel, M., Kalamees, R., Zobel, M., and Rosen, E. 1998. Restoration of species-rich limestone grassland communities from overgrown land: The importance of propagule availability. *Ecological Engineering* 10:275–286.

Pickett, S. T. A. 1989. Space-for-time substitutions as an alternative to long-term studies. In: *Long-term Studies in Ecology.* G. E. Likens (ed.). New York: Springer, pp. 110–135.

Pickett, S. T. A., Collins, S. L., and Armesto, J. J. 1987. A hierarchical consideration of causes and mechanisms of succession. *Vegetatio* 69:109–114.

Powell, E. A. 1992. Life history, reproductive biology, and conservation of the Mauna Kea silversword, *Argyroxiphium sandwicense* DC (Asteraceae), an endangered plant of Hawaii. Ph.D. Dissertation, University of Hawaii, Manoa.

Prach, K. 1994. Succession of woody species in derelict sites in central Europe. *Ecological Engineering* 3:49–56.

Ramensky, L. G. 1924. Basic regularities of vegetation covers and their study. (In Russian) Vestnik Opytnogo dêla Strende-Chernoz. Ob. Voronezh, 37–73.

Reinhart, K. O., Greene, E., and Callaway, R. M. 2005. Effects of *Acer platanoides* invasion on understory plant communities and tree regeneration in the northern Rocky Mountains. *Ecography* 28:573–582.

Robinson, G. R., and Handel, S. N. 2000. Directing spatial patterns of recruitment during an experimental urban woodland reclamation. *Ecological Applications* 10:174–188.

Rui, L., Liu, G., Xie, Y., Qinke, Y., and Liang, Y. 2002. Ecosystem rehabilitation on the Loess Plateau. In: *Regional Water and Soil Assessment for Managing Sustainable Agriculture in China and Australia.* T. R. McVicar, L. Rui, J. Walker, R. W. Fitzpatrick, and L. Changming (eds.). Canberra: Australian Centre for International Agricultural Research, pp. 358–365.

Russell, J. S., and Isbell, R. F. (eds.). 1986. *Australian Soils: The Human Impact.* St. Lucia: University of Queensland Press.

Samuels, C. L., and Drake, J. A. 1997. Divergent perspectives on community convergence. *Trends in Ecology and Evolution* 12:427–432.

Sheley, R. L., and Krueger-Mangold, J. 2003. Principles for restoring invasive plant-infested rangeland. *Weed Science* 51:260–265.

Shiels, A. B., and Walker, L. R. 2003. Bird perches increase forest seeds on Puerto Rican landslides. *Restoration Ecology* 11:457–465.

Shugart, H. H., and West, D.C. 1980. Forest succession modeling. *BioScience* 30:308–313.

Sukopp, H. 2004. Human-caused impact on preserved vegetation. *Landscape and Urban Planning* 68:347–355.

Sukopp, H., and Starfinger, U. 1999. Disturbance in urban ecosystems. In: *Ecosystems of Disturbed Ground*. L. R. Walker (ed.). Amsterdam, The Netherlands: Elsevier, pp. 397–412.

Temperton, V. M., Hobbs, R. J., Nuttle, T., and Halle, S. (eds.). 2004. *Assembly Rules and Restoration Ecology*. Washington, D.C.: Island Press.

van Andel, J., and Aronson, J. (eds.). 2006. *Restoration Ecology*. Oxford, U.K.: Blackwell.

van Diggelen, R. 2006. Landscape: Spatial interactions. In: *Restoration Ecology*. J. van Andel and J. Aronson (ed.). Oxford, U.K.: Blackwell, pp. 31–44.

Wackernagel, M., and Rees, W. 1996. *Our Ecological Footprint: Reducing Human Impact on the Earth*. Gabriola Island: New Society.

Wali, M. K. (ed.). 1992. *Ecosystem Rehabilitation*. The Hague: SPB Academic Press.

Walker, J., and Reuter, D. J. 1996. *Indicators of Catchment Health: A Technical Perspective*. Collingwood: CSIRO Publishing.

Walker, J., Sharpe, P. J. H., Penridge, L. K., and Wu, H. 1989. Ecological field theory: The concept and field tests. *Vegetatio* 83:81–95.

Walker, J., Thompson, C. H., Fergus, I. F., and Tunstall, B. R. 1981. Plant succession and soil development in coastal sand dunes of subtropical eastern Australia. In: *Forest Succession, Concepts and Application*. D. C. West, H. H. Shugart, and D. B. Botkin (eds.). New York: Springer, pp. 107–131.

Walker, J., Thompson, C. H., Reddell, P., and Rapport, D. J. 2001. The importance of landscape age in influencing landscape health. *Ecosystem Health* 7:7–14.

Walker, L. R. 2005. Restoring soil and ecosystem processes. In: *Forest Restoration in Landscapes: Beyond Planting Trees*. S. Mansourian, D. Vallauri, and N. Dudley (eds.). New York: Springer, pp. 192–196.

Walker, L. R., and Chapin, F., S., III. 1987. Interactions among processes controlling successional change. *Oikos* 50:131–135.

Walker, L. R., and del Moral, R. 2003. *Primary Succession and Ecosystem Rehabilitation*. Cambridge: Cambridge University Press.

Walker, L. R., Clarkson, B. D., Silvester, W. B., and Clarkson, B. R. 2003. Colonization dynamics and facilitative impacts of a nitrogen-fixing shrub in primary succession. *Journal of Vegetation Science* 14:277–290.

Walker, L. R., Zarin, D. J., Fetcher, N., Myster, R. W., and Johnson, A. H. 1996. Ecosystem development and plant succession on landslides in the Caribbean. *Biotropica* 28:566–576.

Walker, T. W., and Syers, J. K. 1976. The fate of phosphorus during pedogenesis. *Geoderma* 15:1–19.

Wardle, D. A., Walker, L. R., and Bardgett, R. D. 2004. Ecosystem properties and forest decline in contrasting long-term chronosequences. *Science* 305:509–513.

Warming, E. 1895. *Plantesamfund: Grunträk af den Ökologiska Plantegeografi.* Copenhagen: Philipsen.

Williams, B.G., Walker, J., and Tane, H. 2001. Drier landscapes and rising water tables: An ecohydrological paradox. *Natural Resource Management* 4:10–18.

Wilson, S. D. 1999. Plant interactions during secondary succession. In: *Ecosystems of Disturbed Ground*. L. R. Walker (ed.). Amsterdam, The Netherlands: Elsevier, pp. 611–632.

Young, T. P., Petersen, D. A., and Clary, J. J. 2005. The ecology of restoration: Historical links, emerging issues and unexplored realms. *Ecology Letters* 8:662–673.

Zedler, J. B. 2005. Ecological restoration: Guidance from theory. *San Francisco Estuary and Watershed Science* 3(2):31. http://repositories.cdlib.org/cgi/viewcont.cgi?article=1032&context=jmie/sfews

Zedler, J. B., and Callaway, J. C. 2003. Adaptive restoration: A strategic approach for integrating research into restoration projects. In: *Managing for Healthy Ecosystems.* D. J. Rappaport, W. L. Lasley, D. E. Rolston, N. O. Nielsen, C. O. Qualset, and A. B. Damania (eds.). Boca Raton: Lewis, pp. 167–174.

Zougmore, R., Zida, Z., and Kambou, N. F. 2003. Role of nutrient amendments in the success of half-moon soil and water conservation practice in semiarid Burkina Faso. *Soil and Tillage Research* 71:143–149.

Insights Gained from Succession for the Restoration of Landscape Structure and Function

Roger del Moral, Lawrence R. Walker, and Jan P. Bakker

Key Points

1. The study of succession provides valuable lessons for improving the quality of restoration programs.
2. These lessons suggest that restoration tactics should focus on site amelioration, improving establishment success, and protecting desirable species from herbivory and competition during their development.
3. Incorporation of physical heterogeneity in the early stages will foster mosaics of vegetation that better mimic natural landscapes.

2.1 Introduction

Restoration starts with the desire to improve degraded and destroyed landscapes or ecosystems. Land can be returned to utility through enhancing fertility, by reversing the long-term effects of agriculture, mining, or logging or by ameliorating toxicity. Plant communities also can be modified to resemble their former condition in an effort to provide conservation benefits (van Andel and Aronson 2006). In this chapter, we focus on insights from succession that enhance the rate and quality of restoration. Restoration outcomes are affected by aboveground and belowground processes, but are usually assessed as impacts on aboveground structure and function. We emphasize those processes that can be readily manipulated through a model that features "bottlenecks" to effective restoration. To establish a context for this model, we first discuss concepts central to restoration. Our approach highlights those crucial stages of succession where restoration efforts are most likely to be effective. Here, we highlight how understanding natural succession provides insight into creating effective restoration outcomes. We describe how both structure and function develop during natural succession in response to disturbances. Finally, we summarize the lessons learned from succession that are important in restoration.

2.1.1 Goals

The chances of success in restoration are enhanced if clear goals are established that describe measurable targets to be reached by a specific time. For example, a goal may be to achieve a complex of persistent, species-rich communities with wildlife habitats and opportunities for recreation. This goal could be evaluated by monitoring wildlife populations or plant species and by documenting human usage.

Several strategies might guide a project, but exact mimicry of natural successional trajectories should not be one of them. Because succession is affected by landscape factors and often proceeds slowly, careful intervention usually must occur and the early introduction of target species, those species planned to form the final community, should always be considered. For example, using legumes to enhance soil nitrogen and ameliorate site conditions can foster the development of complex structure decades faster than the direct and continued application of inorganic nitrogen. Restoration actions attempt to guide the trajectory toward desired targets more quickly than would occur spontaneously (cf. Díaz *et al.* 1999).

2.1.2 Ecosystem Parameters

Structure and function are crucial components of ecosystems. The structure of vegetation can be described by species composition (e.g., richness, abundance, dominance hierarchies), by growth form spectra, or by physiognomy. Ecosystem functions include productivity, nutrient cycling, and water use. Species may not contribute to ecosystem function in proportion to their abundance (Schwartz *et al.* 2000) and a few species can dominate functions. These dominant species may usurp resources (luxury consumption) and thereby lower productivity. However, the relationship between dominance and proportional contributions to functions remains debatable. Nearly complete functional restoration often occurs before structure is fully developed, but goals of restoration projects often emphasize structure over function (Lockwood and Pimm 1999). A system may be optimally productive, nutrient conservative, and structurally complex long before it hosts its full complement of species. While increased biodiversity can enhance productivity in grasslands of intermediate fertility, it remains unclear if this effect is proportional to biomass increases (Hector *et al.* 1999, Roscher *et al.* 2005). Plant species are also characterized by adaptive strategies (Grime 2001), in which growth rates and competitive abilities categorize species functions. The mix of strategies found in vegetation shifts during ecosystem development in response to fertility and competition and is therefore sensitive to modification, so trajectories can be under some degree of control. Biodiversity also changes with fertility because both hyper-fertile and infertile sites share low diversity, but have species of contrasting strategies (but see Chapter 6).

Díaz *et al.* (2004) showed that it is possible to predict ecosystem function using simple plant functional traits, so that selecting plants with particular traits can improve these functions. Traits such as leaf size, rooting depth, canopy architecture, seed size, and life span are correlated to productivity and to stress tolerance. By classifying species into functional groups, the task of monitoring ecosystem function is simplified. Even early in succession, these traits track vegetation dynamics. Often, a goal of restoration is to achieve substantial ecosystem structure quickly in order to optimize ecosystem function. Limits to productivity due to infertility and moisture (Baer *et al.* 2004) can retard succession (del Moral and Ellis 2004), so augmenting productivity is often central to restoration. Unfortunately, high productivity often only favors competitive species that produce dense vegetation and arrest structural development and limit biodiversity. Thus, restoration programs in relatively fertile sites, where the priority is to attain high biodiversity quickly, may fail unless fertility is limited and monitored.

2.1.3 Succession and Responses to Environmental Impacts

Succession is the process of species replacements accompanied by ecosystem development. Disturbances cause abrupt changes in or losses of biomass, usually associated with similar changes in ecosystem function. Succession occurs after disturbances that range from mild to severe. Mild disturbances, such as infrequent light ground fire regimes in fire-tolerant vegetation, do little damage. While relative proportions of species change following mild disturbances, species turnover is not directional. Nutrients may be lost and many individuals die, but most species survive. This process of recovery is sometimes called regeneration dynamics, not succession. It is uncommon that restoration will be required in such cases, unless diversity enhancement is required to overcome the consequences of overgrazing or intense fires.

Secondary succession occurs after more severe disturbances such as canopy fire (Beyers 2004) and flooding. Common anthropogenic examples include recovery when farming or grazing cease (Bakker and van Wieren 1998). A legacy of species may persist, but often it consists of undesirable nontarget species. Achieving structure and function comparable to developed vegetation may take decades if the only species that persist are those adapted to disturbances. In these cases, restoration can establish more complex, efficient ecosystems by early, targeted species introductions.

Primary succession occurs after severe disturbances that form new surfaces. Rarely is there a biological legacy, so regeneration is driven from outside the site. Familiar natural examples include lavas, surfaces revealed by retreating glaciers, landslides, and floods (Walker and del Moral 2003). The trajectory of development is unpredictable because the lack of survivors leaves a blank slate upon which many alternatives might be established.

The predictability of restoration can be improved by introducing species expected to form the fundamental structure of the desired system (Turner *et al*. 1998). Definitive model communities for restoration ("nature target types") exist for The Netherlands (Bakker 2005), and could be developed for other regions from available descriptions of plant communities (Rodwell 1991–2000, Schaminée 1995–1999, Wolters *et al*. 2005). Choosing a model community, or suite of communities, depends on the historical context of the target. Restoring rural landscapes to include examples of meadows under moderate grazing, for example, requires data from 19th century descriptions (Bignal and McCracken 1996). However, we emphasize that restoration for biodiversity conservation should aim at multiple targets and a mosaic of habitats. In some cases, no target or model community is known in detail, so target communities must be improvised.

2.1.4 Structure and Function

If an ecosystem has suffered only minor disturbance, structure and function may develop together. Few of the missing elements require immediate replacement because survivors, being physiologically and morphologically plastic, can compensate until others return. Sites that impose physiological stress on plants, such as mine tailings, recover slowly, and have low biodiversity for decades, but functions such as production rates are maximized more quickly than are ecosystem characteristics such as vertical complexity and biodiversity.

Complex structure can be inconsistent with achieving low erosion, high productivity, and tight nutrient cycles in a short time. For example, if most of the species are annuals, much of the surface will be barren during part of the year. High species diversity can be achieved by limiting fertility and competition, but this could reduce productivity and limit nutrient uptake. Much of the literature describing the effects of biodiversity on ecosystem structure concerns species loss, not species additions. Smith and Knapp (2003) showed that net production was scarcely affected by excluding rare species compared to a quantitatively similar reduction of the dominant. However, they suggested that the lack of uncommon species could reduce productivity, thus leading to less inefficient ecosystems. Rosenfeld (2002) suggested that function would be best maintained if the functional group diversity, not species diversity, were maximized. Several biodiversity–ecosystem function experiments in grasslands support this view because the number of functional groups was positively related to ecosystem processes (Hille Ris Lambers *et al.* 2004, Spehn *et al.* 2005).

If functions such as the rate of productivity and nutrient uptake increase more rapidly than does diversity, then restoration can concentrate on dominant species to provide a framework of structure with substantial functioning. This could provide an acceptably stable system with low diversity. Over longer periods, additional species and functional types (*sensu* Díaz *et al.* 1999) can be encouraged to assemble to provide greater long-term resilience.

2.2 Conceptual Scheme of Succession

Natural ecosystem recovery displays an inspiring diversity of responses to equally diverse disturbances. Sites made barren by human activities were once ignored while succession ran its fitful and inefficient course, leading to landscapes replete with exotic species and with limited productivity. The scarcity of productive land and effective biotic reserves now dictates that these sites be restored. Succession provides a framework, not a precise model to enhance restoration efficiency. Restoration often is driven by the real need to achieve effective vegetation cover in a short time. However, where conservation goals are paramount, early successional communities often form a significant component of the resulting landscape. One goal that would mimic nature would be a shifting mosaic of vegetation types that reflect, at any given time, an array of communities attuned to different combinations of fertility, disturbance, and competition, but dominated by native species.

Egler (1954) was among the first to emphasize the vagaries of succession. His initial floristic composition model stated that succession started with whatever propagules were available, even if the species were normally common in late successional stages. Many numerical models emphasize particular aspects of vegetation dynamics, but few usefully predict precise trajectories of all species over long periods (Walker and del Moral 2003). Our model (Fig. 2.1) is not

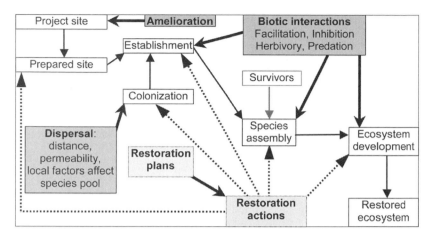

Figure 2.1 Simplified mechanisms of ecosystem change. Restoration plans drive the process and continue to be important throughout the project. The three dark boxes represent natural mechanisms that alter the success of organisms and are linked to processes by thick arrows. Restoration actions and five dotted lines emanating to process boxes indicate restoration actions that may act initially on the project site and subsequently on four phases of restoration. Thin, solid arrows indicate the course of succession. The restoration process starts with planning, in which critical stages are identified. It ends with the formation of the restored ecosystem. Further disturbances, not shown, can affect development at any stage.

comprehensive but does emphasize those constraints that limit and direct species assembly and ecosystem development that can be applied to restoration actions (see Chapter 1).

Three sets of mechanisms direct natural colonization and establishment (shaded boxes, Fig. 2.1): physical amelioration, dispersal, and biotic interactions. After establishment, species form distinct combinations with some becoming dominant as soils develop and biotic interactions intensify (Table 2.1). Restoration actions can alter colonization, establishment, and species accumulations and through these affect ecosystem development. Below we summarize five major phases of succession and suggest how restoration can use this model. At each phase, we first indicate how succession normally occurs and then the relevance for restoration.

2.2.1 Amelioration

Infertility is the most common obstacle to effective restoration. Drought, lack of organic matter, surface instability, and toxicity are among many factors that can also be problematic. These adverse conditions can be created by natural phenomena (e.g., volcanic eruptions, floods) or by human activities (mining, logging). It is rare that destroyed sites will recover both complex structure and substantial function without some applied amelioration (Snyman 2003). During primary succession, natural processes normally improve growing conditions for plants. Winds deposit dust, pollen, seeds, and insects crucial to reducing infertility (Hodgkinson *et al.* 2002). Amelioration can include water erosion that removes overburden (del Moral 1983), frost–thaw cycles that fracture rocks, and wind erosion that creates microtopography to form safe-sites.

Table 2.1 Comparison of lessons from succession and applications to restoration. Topics refer to boxes in Fig. 2.1.

Topic	Lessons from succession	Application to restoration
Amelioration	Stress restricts establishment; safe-site creation important; low fertility may increase diversity	Create heterogeneity and reduce infertility and toxicity
Dispersal	Regional species pool limited; chance is important	Introduce poorly dispersed species in early stages
Colonization	Disharmony characterizes early vegetation; survival probabilities low and stochastic	Introduce array of life-forms; natural dispersal does not provide most colonists; plant more species than required
Establishment	Affected by local variations in stress; oases are of minor importance; safe-sites crucial	Create heterogeneity and safe-sites
Facilitation and inhibition	Nurse plants important; strong dominance reduces diversity; priority effects common	Ameliorate site factors and dominants in mosaic; plant "seral" species at the start to direct trajectory
Herbivory	Animals can eliminate potentially successful species	Protect plantings from large grazers; protect seeds from small seed predators; intermix plantings
Species assembly	Affected by chance, biotic interactions; alternative trajectories are common, sometimes induced by differential herbivory	Accept that there are several viable structural and functional results
Ecosystem development	Strongly affected by biotic interactions, later disturbances	Plan for more disturbance response; manage biotic effects

During restoration, amelioration is usually needed to alter fertility or reduce toxicity. Reid and Naeth (2005) bravely attempted to revegetate mine tailings under subarctic conditions. Kimberlite tailings lack surface stability, organic matter, and nutrients, but do have excessive magnesium from serpentine rocks. By amending soil with organic matter to improve structure and fertility, they established grass cover. However, excessive fertility can reduce biodiversity by promoting only a few competitive species. Biomass responses to fertility are a major control of diversity, at least in grasslands (Grime 2001). Moderate disturbances from mowing or grazing by vertebrates can enhance diversity in more fertile sites by reducing competitive dominance. In some systems, dense, low-diversity vegetation may be desirable to reduce invasion by weeds or to survive intensive use.

Less attention is paid to spatial heterogeneity in physical properties and to variations in fertility, yet these conditions potentially enhance survival of less competitive species and permit sites to develop at different rates. The resulting mosaic enhances overall biodiversity. Huttl and Weber (2001) showed that pine plantations were more successful on coal tailings where acidity varied naturally, providing roots and mycorrhizae opportunities lacking in homogenous acidic soils. Heterogeneity initially present often disappears due to erosion or plant development. Soil heterogeneity in restored prairies near Chicago (USA) declined as C-4 grasses achieved dominance (Lane and Bassiri Rad 2005). Monitoring soil parameters and spatial patterns of dominant species should be included in traditional monitoring, with contingencies to augment heterogeneity if the system becomes too homogeneous. One general method is to import soils with contrasting properties (e.g., acid soils in limestone regions). Alternatively, species that produce litter with qualities distinct from the common species could be introduced.

2.2.2 Dispersal

The ability of most species to disperse is more limited than generally realized, so dispersal can limit colonization (Fuller and del Moral 2003; see Chapter 6). Isolation favors colonization by species with small, buoyant seeds. If the seed rain is sparse, then chance plays a role in species assembly and alternative compositions in similar habitats can develop (McEuen and Curran 2004, Svenning and Wright 2005). Sites that have been severely damaged often have a depleted seed bank with little chance of replenishment (Bakker and Berendse 1999).

Landscape permeability, that feature which resists or promotes dispersal, varies greatly. Permeable landscapes may contain barriers, but also stepping-stones and corridors (Fig. 2.2). Some habitats are impermeable to some species, but not to others (Honnay *et al.* 2002). Barriers and inhospitable habitats reduce permeability and therefore can limit the diversity of functional types that reach a site unaided. Restoration activities can eliminate dispersal problems by planting most species expected in the community. This is rarely successful because residual species resist the newly planted species and swarms of invading alien species can overwhelm the site. Many species that could be effective in a particular project, even though they may not be present in local examples of the target vegetation, are valid candidates for planting. Martínez-Garza and Howe

Figure 2.2 Refugia with shallow pumice depths allowed some species to survive the 1980 eruption of Mount St. Helens (right side of picture). However, the surroundings were impermeable to colonization by the survivors because of deeper pumice deposits. These deposits were colonized by invading species such as *Chamerion angustifolium* shrubs shown in the center of the picture.

(2003) showed that dispersal of rain forest trees into abandoned pastures was severely limited and that planting trees shortened succession by at least three decades.

The restoration of diverse meadows from pastures is particularly difficult when the existing ruderals resist the establishment of meadow species. In such cases, dispersal can be promoted by the introduction of hay from a reference site (Hölzel and Otte 2003) and by adding top soil and litter with seeds of target species and appropriate soil organisms (van der Heijden *et al.* 1998, De Deyn *et al.* 2003).

2.2.3 Colonization

Isolated sites are unlikely to receive large-seeded species common in later succession, so early communities are a disharmonious selection of the local flora. Species that do arrive are usually small-seeded and without large energy reserves. While a few individuals may establish, early development is commonly limited to very favorable sites (see also Wagner 2004). Seedling failure rates are also high. When understory species were planted in *Fagus* forests in Belgium, survival was higher in cleared sites than in the controls (Verheyen and Hermy 2004). Colonization by *Pinus sylvestris* in Spanish old-fields was restricted both by competition from meadow vegetation and by seed predation (Castro *et al.* 2002). Such failures to establish slow the rate of ecosystem development.

There is a temptation to depend on spontaneous recolonization because it is economical (see Chapter 6). Prach *et al.* (2001a) suggested that spontaneous succession (i.e., depending on volunteering colonists) might be useful, at least for reclamation, where any vegetation at all is beneficial. Ideally, spontaneous species will facilitate the establishment of woody species in forest environments, but this is uncertain. The nature of volunteer species is contingent on the landscape, and trajectories started by ruderal species often diverge in unexpected ways and lead to vegetation that provides few values (Prach *et al.* 2001b). Unless economic resources available for restoration are scarce, even a favorable seed rain of spontaneous species should not preclude the introduction of target species. Under vegetation conditions typical of restoration programs, where the surroundings are disturbed and mature vegetation is scarce, spontaneous vegetation often will be dominated by exotic species (Bakker and Wilson 2004) and active introduction of species will be required when biodiversity conservation is a goal. There are several ways to enhance the colonization of spontaneous species, though each has limitations. Installing perches creates centers of dispersal for species dispersed by birds (Toh *et al.* 1999), but better still is to use trees and shrubs that attract birds and that can protect seedlings (Slocum and Horvitz 2000). This nucleation process is a crucial form of colonization in many types of natural succession and can accelerate the establishment of desirable species.

2.2.4 Establishment

The establishment phase is critical, and surfaces can be hostile. A seedling must grow rapidly to reach better conditions, a feat constrained by infertility, drought, excessive light, surface heat, and other unfavorable conditions. Establishment is promoted by mechanisms that trap seeds to increase the odds of germination, by stable surfaces, and by safe-sites appropriate to each species (Walker *et al.*

2006). Jones and del Moral (2005) noted that seedlings on a glacial foreland were normally found in microsites that offered substantial protection, while Tsuyuzaki *et al.* (1997) showed that even minimal surface instability restricted seedling establishment on loose volcanic substrates.

Establishment success may be improved if several species in each of several functional groups (functional redundancy) are employed early in restoration. Even if some species fail, ecosystem functions are likely to develop more quickly than if too much reliance is placed on a few species. Using functional redundancy may prove beneficial in view of unpredictable global change.

2.2.4.1 Facilitation

Biological facilitation is the process by which established plants improve the performance of other plants. Facilitation has been the process emphasized during establishment, largely because of early studies of succession that were focused on this process (Walker and del Moral 2003). Facilitation is physical when established plants improve soil moisture availability, temperature, or light conditions or reduce wind. Rocks and small channels augmented seedling survival early in succession on Mount St. Helens (del Moral and Wood 1993; Fig. 2.3), but plants also provide physical amelioration (Barchuk *et al.* 2005). Established plants may be "nurse plants" and facilitate seedling establishment (Henríquez and Lusk 2005). Nurse plants may inhibit one species, thus releasing other species from competition. Legumes are particularly likely to have such complex interactions (del Moral and Rozzell 2005). Eventually, the fostered plant may eliminate the nurse plant (Temperton and Zirr 2004). Shrubs often

Figure 2.3 *Anaphalis margaritacea* is one of several species that were able to establish early in succession, following the 1980 eruption of Mount St. Helens, by lodging among rocks. Rocks, as well as other microsite features, enhance moisture and nutrients, while protecting seedlings from herbivory.

protect forbs from herbivory by physical (Garcia and Obeso 2003) or by chemical means (Jones *et al.* 2003), while late successional species such as *Quercus robur* can establish among spiny *Prunus spinosa* (Bakker *et al.* 2004). However, facilitation should be used carefully so that it does not become inhibition.

Physical amelioration tactics are well-known. In addition, site heterogeneity should be enhanced to improve the number and variety of safe-sites. Rocks, hummocks, and rills foster heterogeneity, provide refuges, and help to insure against unforeseen events. Even small variations on hostile surfaces can facilitate seedling establishment, so small restoration efforts create favorable microsites. Existing heterogeneity on relatively level terrain should be preserved and natural processes mimicked. Heterogeneity can be augmented by the introduction of rocks large enough to protect seedlings from drought or herbivory and by introducing inorganic mulch of variable depths. The creative use of low windbreaks can conserve moisture and reduce desiccation of seedlings. A mosaic imposed at the start of a project, for example by patches with different fertilization regimes, may reduce the need for intense, long-term maintenance. Physical amelioration aimed at creating heterogeneous conditions can lead to the development of alternative, but stable and desirable communities.

2.2.4.2 Inhibition

The inhibitory potential of plants during succession is crucial but little appreciated. Such negative effects of one species on another can slow, arrest, or deflect succession. Competition for resources and allelopathy are the main types of inhibition. Although nitrogen-fixing plants may ultimately facilitate other species, their immediate effect can be inhibitory, particularly when a dense sward or thicket forms. The facilitative effect can be delayed until the nitrogen-fixer dies (Gosling 2005). Such an inhibitory effect of nitrogen fixing plants may be more common than generally realized (Walker 1993). Inhibition often causes problems during restoration. Aggressive invaders suppress plantings or nurse plants suppress desired target species. By planting saplings in scattered clusters to provide mutual support, followed by selective thinning, the growth of species expected to form the framework of mature vegetation can be accelerated. Selective thinning of nurse plants and competitors can also facilitate the development of target species (Sekura *et al.* 2005).

2.2.4.3 Herbivory

Seed predation and other forms of herbivory can reduce establishment (Ramsey and Wilson 1997). However, herbivory can also promote seed dispersal, add nutrients, and facilitate seedling recruitment (Bakker and Olff 2003). Such interactions have been observed to involve livestock, burrowing mammals, and ungulates such as the North American elk on Mount St. Helens. Herbivory during establishment is a major cause of restoration failure. In many cases, plantings must be protected from herbivores by fences or individual exclosures until they become established.

Plant defense against herbivory, such as wood, terpenes, and tannins, generally increases during succession as a function of changing species composition and maturation of individuals. In secondary succession, N-based secondary compounds may defend forbs so herbivory is concentrated on grasses, deciduous shrubs, and trees (Davidson 1993). Because palatable plants often dominate intermediate successional stages, herbivory can retard or expedite succession. Each situation requires analysis to determine whether herbivores should be excluded, at least during crucial phases of the project. Excluding herbivores from parts of a project, but not others, could facilitate a desirable vegetation mosaic.

Herbivores can disrupt dominance and thus permit establishment of new species (Bach 2001). Bishop *et al.* (2005) demonstrated that various herbivores could reduce the rate of succession by slowing the rate of *Lupinus lepidus* expansion. However, herbivory also reduces competition by *Lupinus* and hastens the development of sites in which it had dominated. Herbivory may accelerate succession because some plant species may exhibit compensatory growth in the face of herbivory (Vail 1992, Hawkes and Sullivan 2001). While herbivory is more likely to spawn negative effects (e.g., promoting weed invasion or accelerating erosion), the possibility that it may be positive should be considered for each study (Belsky 1992). Established communities may be changed in unpredictable ways because the conditions of the restoration site may not have a comparable natural model. Howe and Lane (2004) established wetland prairies, and then exposed them to herbivory by voles. Voles caused dramatically divergent trajectories after four years, fostering a mosaic. Ants can hoard certain species to enhance the vegetation mosaic (Gorb *et al.* 2000, Dostal 2005).

2.2.5 Assembly and Ecosystem Development

Species can accumulate over decades, even while successive waves of pioneers fail. Populations expand and fill available space, thus increasing the use of resources. During this time, the character of the community emerges. Planned actions or responses to unexpected contingencies can lead to results that are more desirable, yet little attention has been paid to modifications during species assembly. It is during this period that adequate results can be sharply improved.

2.2.5.1 *Biotic Effects*

The composition of a developing community can be affected by the arrival of additional species, and, as is the case with establishment, by facilitation, inhibition, and herbivory. Facilitation and inhibition continue to have multiple effects (Fig. 2.4). For example, N-fixing taxa improve soil fertility, but they have complex interactions with other plants and with suites of herbivores as well (Bishop 2002, Clarkson *et al.* 2002). The competitive effects of these species often filter the species that could benefit from improved soil fertility. *Lupinus lepidus* initially formed sporadic dense colonies on Mount St. Helens. Because this species is short lived and susceptible to multiple attacks from herbivores, the colonies expanded slowly, and their abundance cycled greatly. After several cycles, species able to establish during "crash" years have become abundant, but mosses make it difficult for other species to establish (del Moral and Rozzell 2005). The totality of how a species affects others, not just its net effect on the community, is crucial to understanding probable trajectories.

As species assemble, competitive hierarchies form and structure develops, but the overall net effects of competition and facilitation are difficult to predict. Hence, the trajectory of the community is also hard to predict. Species composition will adjust over time and usually lead to a functionally integrated

Figure 2.4 *Lupinus lepidus* and mosses interact to form a dense carpet on lahars at Mount St. Helens. Their net effects on other species are complex. While lupines enhanced nitrogen levels, the primary beneficiaries were mosses. Mosses inhibit the establishment of seed plants, while lupines competed with germinating seedlings.

ecosystem with substantial complexity and spatial variation. However, we note exceptions. If dense vegetation becomes established, subsequent development may be arrested. Dense growths of grasses, vines, ferns, bamboo, and shrubs can form thickets that defy change (Walker and del Moral 2003, Temperton and Zirr 2004). Thickets may be useful to restoration if they curtail erosion, reduce herbivory, or improve fertility and if the thicket eventually senesces. Artificial thickets can be created using dead branches to protect young plants. Planting late successional species in dense arrays can enhance their survival, promote heterogeneity, and limit weeds. If the goal is to produce low-maintenance vegetation dominated by shrubs, then shrub thickets can arrest succession (Niering and Egler 1955, Fike and Niering 1999). A mixture of species is usually superior to one because several species complement one another and provide more resources for wildlife (De Blois *et al.* 2004).

Nontarget species can be resisted by proper maintenance of existing target species. Using unpalatable species as "nurse plants" should be considered where herbivory is likely to reduce recruitment or harm young planted species. Callaway *et al.* (2005) described how unpalatable species produced indirect facilitation effects on palatable grassland species in the Caucasus (Russia). The benefits of using indirect facilitation include greater biological and functional diversity, though care must be exercised that the facilitators do not become dominant.

The difficult balance among biotic effects is illustrated by attempts to enhance the biodiversity of abandoned grasslands. Grazing and biomass removal is often insufficient to reduce fertility, a prerequisite to promoting higher diversity of target species. Topsoil removal (see Chapter 6) is a viable tactic, but nontarget species often invade and dominate the disturbed conditions. Sowing pasture grasses in an effort to smother nontarget species often creates a dense turf that inhibits target species (Bakker 2005).

2.2.5.2 Further Disturbances

Restoration projects are not immune from further disturbances due to grazing, fire, wind, or disease. Most disturbances are ephemeral, but some, such as herbivory, can destroy a project. Therefore, even after establishment, young plants often require protection, and exclosures against large animals are frequently needed (Koch *et al.* 2004).

Fire can destroy a restoration project, but often it merely serves to rejuvenate the vegetation and to promote the growth of target species. Frequent fires usually create herbaceous vegetation dominated by short-lived species, while less frequent fires can promote fire resistant trees (Hooper *et al.* 2004). Using fire to introduce or maintain heterogeneity can promote diversity at the scale of the project.

Atmospheric nitrogen deposits are a major disturbance that affects the structure and function of all ecosystems. Greater fertility lowers diversity by favoring only a few competitive species (Zavaleta *et al.* 2003, Suding *et al.* 2005). This effect is widespread, affecting not only industrialized regions, but also such isolated areas as the Mojave Desert, California, where atmospheric nitrogen deposits promoted alien plants and inhibited native species (Brooks 2003). Nitrogen deposition can also facilitate shifts in vegetation types (Kochy and Wilson 2005).

One approach to reduce excess fertility is to remove biomass. After long-term haymaking without fertilization, the output through hay was higher than the input from atmospheric nitrogen deposition. The critical input to maintain nutrient poor meadow communities in northern Europe is less than atmospheric deposition, suggesting that no fertilization is needed in these habitats (Bobbink *et al.* 1998).

2.2.5.3 Restoration Disturbances

Restoration is a unique form of disturbance, applied over time in a nuanced way. Adding fertility in a mosaic, for example, is a disturbance because it alters the existing regime in ways designed to alter species composition. The desired result is a vegetation mosaic with horizontal and vertical heterogeneity, even in grasslands and subalpine vegetation. Restoration tactics may deflect the trajectory of an assembling community in several ways. Species exerting strong dominance may be thinned. Fires or secondary disturbances may be introduced and, at some stage, it may be imperative to introduce mycorrhizae to foster more complete ecosystem function (Allen *et al.* 2005). There are many opportunities to introduce integral species incapable of independent establishment. Zanini and Ganade (2005) showed that perches that attracted birds to abandoned Brazilian subtropical pastures enhanced diversity of woody species. More seedlings were introduced where residual vegetation occurred, suggesting that a facilitative effect was also present. White *et al.* (2004) confirmed that spontaneous establishment is unreliable. In North Queensland forests, dispersal into isolated revegetated forest remnants was inundated by exotic species. Humans must intervene to introduce species into isolated recovering sites (Holl *et al.* 2000). Where restoration efforts occur on small sites, attracting bird dispersers may have little effect because the seed rain is dominated by wind-dispersed species that inhibit the few woody seedlings (Shiels and Walker 2003).

Relict and rapidly establishing vegetation present major challenges to restoration. Management is needed to overcome the inertia of survivors and exotic invaders. Hooper *et al.* (2005) demonstrated many barriers to regeneration of tropical forests on abandoned pastures in Panama. Competition from grasses, limited seed dispersal, and fire all restricted potential colonists. By planting native species in clusters, providing firebreaks, and abstaining from fertilization, recovery was promoted.

2.2.5.4 Community Effects

Successful restoration requires an understanding of individual species, but relatively early in the process the focus must shift to community effects. Communities form as species proportions shift through competition and facilitation, colonization by species, and differential herbivore and disease pressures. Competition and facilitation vary in space and time, depending on the density of the participant species. While a canopy species can provide understory heterogeneity, biodiversity often declines (Morgantini and Kansas 2003). The results of the complex biotic interactions include divergent trajectories to both undesirable states that need to be redirected and to acceptable communities. The rates by which species facilitate or inhibit others differ with environmental stress, so succession rates will differ locally to create biologically heterogeneous conditions. By altering stress levels, desirable heterogeneity in a restoration project can be promoted. This heterogeneity can provide shifting spatial conditions so that no species achieves strong dominance during assembly. Once initiated, heterogeneity persists and provides greater structural complexity.

Restoration projects that are impacted by severe disturbances may not be able to recover the spectrum of species types found in mature, intact vegetation (Dana *et al.* 2002), or even recover their pre-disturbance functions. Dispersal limitations and local depletion of biodiversity can preclude many species from colonizing (Pyšek *et al.* 2005), so ongoing management could promote species with limited dispersal or reestablishment difficulties. If a project develops only from common species, structure will suffer and functions may be suppressed. Where the landscape matrix is agricultural, the promotion of species complexity may be more important as one way to provide habitats for species that can limit agricultural pests.

Under many conditions, restoration will be successful if there is complex growth-form structure with desired target species, even if biodiversity remains low. Over time, greater biotic complexity may accumulate, but it is likely that it will be much less than natural vegetation (Rayfield *et al.* 2005). During development, monitoring should continue to determine if interventions are needed. It is rare that they are not. Dominance by a few thriving species or invasion of nontarget species requires attention. Disturbances from grazing, fire, pathogens, or wind all may require attention. Monitoring is also required to note the need to intervene to nudge the system along more desirable trajectories (de Souza and Batista 2004). In some cases, low biodiversity is acceptable because it reflects the natural situation (e.g., a salt marsh or a fen) and is the desired target. In other cases, limited biodiversity is an adequate result if community processes are adequate and the goal is to alleviate erosion or provide amenities.

2.3 Restoration Planning

Planning to facilitate the recovery of a landscape from anthropogenic impacts requires knowledge of the site, of potential ecosystems that can be achieved, and of the bottlenecks to development (Temperton *et al.* 2004, van Andel and Aronson 2006). A clear idea of the nature of the site when active maintenance ceases should be part of any plan. Planning not only prescribes the procedures and protocols, but also provides for maintenance and management to reach specific goals. It specifies the criteria by which a project is evaluated. Effective planning includes proper monitoring that will be communicated in the open literature. In this way, effective methods will be disseminated and mistakes can be avoided. Restoration should focus on five stages (Fig. 2.1), though for practical reasons, most effort will be put on amelioration of the environment and establishment. Colonization occurs *de facto* when species are selected, but many programs ignore species assembly and ecosystem development.

Planning starts with goals. Because late successional vegetation under similar environments can be variable (McCune and Allen 1985) and because trajectories are unlikely to converge to predictable endpoints (Taverna *et al.* 2005), goals should be specified in functional terms after considering the landscape and its biota (Khater *et al.* 2003). Functional goals can reside within goals expressed as structural classes such as short swards or tall forb communities and their spatial arrangement (Bakker 1998). Biodiversity goals derived from community descriptions are available in many countries (e.g., Anderson 2005). The selected species should be capable of forming a functional community, and their life-history characteristics can be incorporated into planning (Knevel *et al.* 2003).

Before the start of major projects, existing soil conditions (e.g., fertility, moisture, microsites), surviving species (if any), and local topography must be determined. These parameters will help limit the range of feasible "targets." During planning, pilot studies with bioassay species (e.g., fast growing grasses) can help determine needs for site amelioration. In extreme cases, bioremediation may be required to reduce toxicity. At the same time, the ability of dominant species to establish under planned amelioration tactics should be determined in field trials (Palmer and Chadwick 1985). Pilot studies and field trials will provide a substantial return on their investment and significantly increase the probability of success.

Contingency planning requires a pessimistic view and a willingness to consider rescue programs. Potential problems are associated with competition, infertility, and herbivory. The competitive environment must be assessed. Plans to remove exotic and nontarget species and to thin target species should be in place, with specific triggers in the maintenance plans (Ogden and Rejmanek 2005). Fertility often limits development when initial stores of nutrients become sequestered in the standing vegetation (Feldpausch *et al.* 2004), so nutrient stress should be monitored. Other common problems, such as episodic herbivore damage, catastrophic weather events, and unforeseen changes in the local environment all need to be addressed.

2.4 Lessons from Succession

Effective ecological restoration of barren, derelict, and degraded landscapes requires attention to the messages produced by natural recovery of ecosystems. Restoration often involves sites without vegetation or those dominated by non-target species that are isolated from pools of natural colonists. Here, restoration starts with alteration of abiotic conditions. Other sites, however, require enhancements of their properties. Heterogeneity may be reintroduced, erosion and sedimentation controlled, and competition limited by grazing, mowing, or topsoil or sod removal. Succession is not the predictable process it was once believed to be. It requires dynamic management at each stage because of this unpredictability and multiple outcomes should be accepted, if not always entirely welcomed.

2.4.1 Restoration Phases

There are three major phases in the redevelopment of a community (Table 2.2). A major goal is to enhance the structure and function of the site to improve ecosystem health (Cramer and Hobbs 2002). Healthy systems are resistant to further impacts, experience only limited fluctuations in population numbers, and are productive. The type of enhancement is determined in part by local circumstances (Bakker and Londo 1998). For example, the desired level of biodiversity may be lower in an industrial park compared to a rural area. However, the tactics differ in each of the stages. Environmental restoration is sometimes appropriate in the aftermath of major natural disturbances (e.g., lahars) that create new surfaces, but it is more common in intensively affected cultural landscapes (e.g., mine wastes). Physical amelioration, such as erosion control, and species introductions dominate this phase of restoration as the community is directed toward defined targets. In degraded cultural landscapes, vegetation is dominated by ruderal species and turnover is rapid. These ruderal species have little conservation interest and little direct economic value, so they should be controlled.

Table 2.2 Characteristics of managed landscapes during community development [modified from Bakker and Londo (1998)]

Characteristics	Early stages	Developing stages	Late stages
Dominating processes	Environmental restoration	Biotic restoration	Maintenance
Biotic function	Low	Moderate, directed	High, maintained; heterogeneous
Biotic structure	Variable, not desired	Increasing, directed	High, heterogeneous
Strategies	Physical amelioration; species introductions	Manage biotic environment	Limited management of populations, environment
Examples	Restore topographic heterogeneity; amend fertility; introduce targets	Selective thinning; grazing regime fits target; limit competition	Replace failed target species; suppress nontarget species
Species characteristics	Ruderal	Competitive, mixture of subordinate species	Competitive, with stress-tolerant species; mixture of subordinate species
Community turnover	High; directed toward multiple targets	Declines as targets are approached	Low, with minor, turnover

During the second phase of recovery, vegetation is often actively managed to improve its conservation value. Additional target species may be introduced, though many can survive early introductions. Many species are competitive, so thinning or mowing may be needed to enhance biodiversity (Bakker *et al.* 2002). Nontarget species should be controlled so that trajectories are directed toward stated targets. Species turnover declines as the vegetation attains maturity. Finally, as the conservation interest of the vegetation is maximized, management becomes focused on maintenance. Monitoring directs management to maintain biodiversity through tactics such as thinning the canopy, reintroducing species that may have died out, and litter removal, leading to a vegetation mosaic. The final community may change cyclically both in space and time and species populations will fluctuate, but turnover is low.

2.4.2 Heterogeneity

Even barren sites may have some desirable heterogeneity. Surviving physical heterogeneity should be preserved and incorporated into plans instead of being graded to uniformity. This may preserve safe-sites, foster biodiversity, and facilitate development of the ecosystem. Variation can be a hedge against the unexpected and can offer a refuge during times of extreme climate. Using several growth forms helps to ensure that extreme events will not destroy all species. Structural variation provides resilience by permitting cores of survivors even if catastrophes occur.

2.4.3 Landscape Effects

The surroundings are nearly as important as the characteristics of the site. They contribute propagules that could augment or inhibit restoration, so their net effects must be considered. Dispersal is inherently subject to chance, so the pool of potential colonists in fragmented landscapes may be drastically different from that of intact landscapes. Target species may be missing or isolated and their low probability of colonization can produce unpredictable results. Restoration must introduce target species at the correct time.

Complex vegetation requires a certain minimum area, and small sites are influenced by the invasion of dispersible species. The effects of the species-area curve have been documented for urban fragments (Murakami *et al.* 2005) and forests (Ross *et al.* 2002). Small sites lose species rapidly and never accumulate a full complement of species (Bastin and Thomas 1999). This suggests that planners should have expectations for complexity based on the size of a restoration project and its surroundings (Margules and Pressey 2000) not on large natural reference areas.

Which species reaches a site is one of the least predictable events. These pioneers can dictate subsequent development by altering soils, and possibly deflecting trajectories (Temperton and Zirr 2004). It is common for different trajectories to develop on the same site due to priority effects, that is, the impact of the first wave of colonists on later arrivals. Because colonization is episodic, initial natural succession is highly variable. Both spatial and temporal variation may be desirable for the development of the ecosystem, so planning should provide for such individualistic results and vegetation mosaics.

One consequence of priority effects and habitat heterogeneity is the development of a mosaic of alternative states, stable yet distinct vegetation types growing together under similar environments. Stochastic processes, differential rates of development, a shifting balance between facilitation and inhibition and secondary disturbances all foster mosaics. Examples are common in riparian vegetation (Baker and Walford 1995) where mature vegetation often exhibits contrasting composition (Honnay *et al*. 2001) and on broad plains recently freed from flooding (del Moral and Lacher 2005). Mosaics augment biodiversity and promote wildlife. A mosaic of types has several virtues, so a variety of targets is often warranted. Biodiversity is enhanced locally through the rescue effect (Gotelli 1991, Piessens *et al*. 2004) where colonists from other patches save target populations from going extinct and on a larger scale by differences among the mosaic elements. Multiple simultaneous trajectories are one way to insure against unforeseen consequences.

2.5 Conclusions

Fully applying the lessons of succession will improve the efficiency and quality of restoration programs by assuring that both structure and function develop well. It is difficult to predict restoration trajectories *a priori* by reference to "assembly rules" derived from species characteristics or studies under different conditions because young plants, planted sparsely, often lack a competitive environment. Studies that do demonstrate assembly rules typically are in competitive environments (Weiher and Keddy 1995, Bell 2005, Fukami *et al*. 2005). Rules can work at the level of functional traits and dispersal types, but are confounded by chance, competition from nontarget species, and stressful conditions (Walker *et al*. 2006). Facilitation and inhibition by the same species is complex and dynamic, so that predicting patterns may require detailed knowledge of local conditions. Natural vegetation is the result of many contingent and stochastic factors so that existing mature vegetation is either only one of several viable alternatives or it is a mosaic. Thus, local mature vegetation may be a guide for planning, but not a detailed model. This is pragmatic because it permits several acceptable species compositions. The benefits of a community with several growth forms (or functional types) with multiple representatives of each may include greater productivity (Hille Ris Lambers *et al*. 2004), resistance to invasion (Symstad and Tilman 2001, Fargione and Tilman 2005), and enhanced ecosystem functions (Symstad *et al*. 2003) compared to a homogeneous site.

The lessons of natural succession provide guidelines even if rules are contingent. Fragmentation, barriers, differential permeability, and isolation filter potential colonists, so that spontaneous recruitment rarely leads to an ecosystem with optimal structure and function. Further, the suite of first colonists will not represent the total pool. Even when economic constraints require dependence on spontaneous recruitment, amelioration helps to select for more desirable species, and improves both the diversity and the density of colonists. Amelioration actions should produce variable substrates that will allow complex vegetation. Mosaics of communities usually characterize early succession. Homogeneous vegetation that results from application of similar procedures and vegetation throughout the project is better replaced by more nuanced actions designed to foster vegetation mosaics.

During the assembly of vegetation, conditions that filter immigrants change, leading to a different set of new colonists. At the least, moisture, nutrients, light, and biotic pressures change, altering the success of existing and immigrant species (Fattorini and Halle 2004). Diversity can be enhanced by the reduction of competition (Polley *et al.* 2005). One way to limit competitive dominance is to plant the less competitive species before putative dominants and to increase the number of species and functional groups early in the restoration process. Though it is appealing to mimic natural succession, planting sequences do not have to follow natural sequences. In nature, many species do not establish early in a trajectory either because they fail to arrive, or having reached the site, cannot establish. During restoration, species can be introduced early in the sequence if the conditions can be manipulated to help them establish. Slower growing species common in stable vegetation can be planted early in the process, in masses, to prevent them from being smothered by other species. This has the added benefit of enhancing the mosaic. Other treatments, including thinning and selective disturbances, may be feasible.

The use of herbivores to facilitate succession is poorly studied, though we know that moderate grazing can sometimes promote diversity. More often, restoration projects must be protected from vertebrates, and sometimes from invertebrates. Intermixing species can slow selective grazers and diverse plantings have other virtues.

Because the species composition of restoration projects can develop in unpredictable ways, composition alone is not the best measure of success. Rather, performance standards might be measured in terms of spatial mosaics, vertical complexity, overall diversity, and reproductive success among the shorter-lived species. Ideally, functions such as biomass accumulation rates and biofiltration efficiency can be used to measure performance.

There is much to be learned from succession. Restoration can help to improve the understanding of succession by monitoring and reporting the results of the application of succession theory (Young *et al.* 2005, see Chapter 1). At the same time, attention to the lessons learned from studies of succession will improve the quality, efficiency, and success of restoration.

Acknowledgments: Roger del Moral thanks the U.S. National Science Foundation for support of his studies on Mount St. Helens (DEB-00-87040). We thank Joseph Antos, Michael Fleming, Ari Jumpponen, Felix Mueller, Rachel Sewell Nesteruk, and Vicky Temperton for careful reviews of the manuscript. Lawrence Walker was supported by a sabbatical from the University of Nevada Las Vegas and by Landcare Research, New Zealand.

References

Allen, M. F., Allen, E. B., and Gomez-Pompa, A. 2005. Effects of mycorrhizae and nontarget organisms on restoration of a seasonal tropical forest in Quintana Roo, Mexico: Factors limiting tree establishment. *Restoration Ecology* 13:325–333.

Anderson, M. 2005. Vegbank, on-line vegetation data bank. URL: http://vegbank.org/vegbank/index.jsp.

Bach, C. E. 2001. Long-term effects of insect herbivory and sand accretion on plant succession on sand dunes. *Ecology* 82:1401–1416.

Baer, S. G., Blair, J. M., Collins, S. L., and Knapp, A. K. 2004. Plant community responses to resources availability and heterogeneity during restoration. *Oecologia* 139:617–629.

Baker, W. L., and Walford, G. M. 1995. Multiple stable states and models of riparian vegetation succession on the Animas River, Colorado. *Annals of the Association of American Geographers* 85:320–338.

Bakker, E. S., and Olff, H. 2003. Impact of different-sized herbivores on recruitment opportunities for subordinate herbs in grasslands. *Journal of Vegetation Science* 14:465–474.

Bakker, E. S., Olff, H., Vandenberghe, C., De Mayer, K., Smit, R., and Gleichman, J. M. 2004. Ecological anachronisms in the recruitment of temperate light-demanding tree species in wooded pastures. *Journal of Applied Ecology* 41:571–582.

Bakker, J. D., and Wilson, S. D. 2004. Using ecological restoration to constrain biological invasion. *Journal of Applied Ecology* 41:1058–1064.

Bakker, J. P. 1998. The impact of grazing on plant communities. In: *Grazing and Conservation Management*. M. F. Wallis DeVries, J. P. Bakker, and S. E. Van Vieren (eds.). Dordrecht: Kluwer, pp. 137–184.

Bakker, J. P. 2005. Vegetation conservation, management and restoration. In: *Vegetation Ecology*. E. van der Maarel (ed.). Oxford: Blackwell, pp. 309–331.

Bakker, J. P., and Berendse, F. 1999. Constraints in the restoration of ecological diversity in grassland and heathland communities. *Trends in Ecology and Evolution* 14:63–68.

Bakker, J. P., and Londo, G. 1998. Grazing for conservation management in historical perspective. In: *Grazing and Conservation Management*. M. F. Wallis DeVries, J. P. Bakker, and S. E. Van Vieren (eds.). Dordrecht: Kluwer, pp. 23–54.

Bakker, J. P., and van Wieren, S. E. (eds.). 1998. *Grazing and Conservation Management*. Dordrect: Kluwer.

Bakker, J. P., Elzinga, J., and De Vries, Y. 2002. Effects of long-term cutting in a grassland system: Perspectives for restoration of plant communities on nutrient-poor soils. *Applied Vegetation Science* 5:107–120.

Barchuk, A. H., Valiente-Banuet, A., and Díaz, M. P. 2005. Effect of shrubs and seasonal variability of rainfall on the establishment of *Aspidosperma quebracho-blanco* in two edaphically contrasting environments. *Austral Ecology* 30:695–705.

Bastin, L., and Thomas, C. D. 1999. The distribution of plant species in urban vegetation fragments. *Landscape Ecology* 14:493–507.

Bell, G. 2005. The co-distribution of species in relation to the neutral theory of community ecology. *Ecology* 86:1757–1770.

Belsky, J. A. 1992. Effects of grazing, competition, disturbance and fire on species composition and diversity in grassland communities. *Journal of Vegetation Science* 3:187–200.

Beyers, J. L. 2004. Postfire seedling for erosion control: Effectiveness and impacts on native plant communities. *Conservation Biology* 18:947–956.

Bignal, E. M., and McCracken, D. I. 1996. Low-intensity farming systems in the conservation of the countryside. *Journal of Applied Ecology* 33:413–424.

Bishop, J. G. 2002. Early primary succession on Mount St. Helens: The impact of insect herbivores on colonizing lupines. *Ecology* 83:191–202.

Bishop, J. G., Fagan, W. F., Schade, J. D., and Crisafulli, C. M. 2005. Causes and consequences of herbivory on prairie lupine (*Lupinus lepidus*) in early primary succession. In: *Ecological Responses to the 1980 Eruption of Mount St. Helens*. V. H. Dale, F. J. Swanson, and C. M. Crisafulli (eds.). New York: Springer, pp. 151–161.

Bobbink, R., Hornung, M, and Roelfos, J. G. M. 1998. The effects of air-borne nitrogen pollutants on species diversity in natural and semi-natural vegetation: A review. *Journal of Ecology* 86:717–738.

Brooks, M. L. 2003. Effects of increased soil nitrogen on the dominance of alien annual plants in the Mojave Desert. *Journal of Applied Ecology* 40:344–353.

Callaway, R. M., Kidodze, D., Chiboshvili, M. and Khetsuriani, L. 2005. Unpalatable plants protect neighbors from grazing and increase plant community diversity. *Ecology* 86:1856–1862.

Castro, J., Zamora, R., and Hodar, J. A. 2002. Mechanisms blocking *Pinus sylvestris* colonization of Mediterranean mountain meadows. *Journal of Vegetation Science* 13:725–731.

Clarkson B. R., Walker, L. R., Clarkson, B. D., and Silvester, W. B. 2002. Effect of *Coriaria arborea* on seed banks during primary succession on Mt. Tarawera, New Zealand. *New Zealand Journal of Botany* 40:629–638.

Cramer, V. A., and Hobbs, R. J. 2002. Ecological consequences of altered hydrological regimes in fragmented ecosystems in southern Australia: Impacts and possible management response. *Austral Ecology* 27:546–564.

Dana, E. D., Vivas, S., and Mota, J. F. 2002. Urban vegetation of Almeria City—A contribution to urban ecology in Spain. *Landscape and Urban Planning* 59:203–216.

Davidson, D. W. 1993. The effects of herbivory and granivory on terrestrial plant succession. *Oikos* 68:23–25.

De Blois, S., Brisson, J., and Bouchard, A. 2004. Herbaceous covers to control tree invasion in rights-of-way: Ecological concepts and applications. *Environmental Management* 33:506–619.

De Deyn, G. B., Raaijmakers, C. E., Zoomer, H. R., Berg, M. P., de Ruiter, P. C., Verhoef, H. A., Bezerer, T. M., and van der Putten, W. H. 2003. Soil invertebrate fauna enhances grassland succession and diversity. *Nature* 422:711–713.

del Moral, R. 1983. Initial recovery of subalpine vegetation on Mount St. Helens, Washington. *American Midland Naturalist* 109:72–80.

del Moral, R., and Ellis, E. E. 2004. Gradients in heterogeneity and structure on lahars, Mount St. Helens, Washington, USA. *Plant Ecology* 175:273–286.

del Moral, R., and Lacher, I. L. 2005. Vegetation patterns 25 years after the eruption of Mount St. Helens, Washington. *American Journal of Botany* 92:1948–1956.

del Moral, R., and Rozzell, L. R. 2005. Long-term effects of *Lupinus lepidus* on vegetation dynamics at Mount St. Helens. *Plant Ecology* 182:203–215.

del Moral, R., and Wood, D. M. 1993. Understanding dynamics of early succession on Mount St. Helens. *Journal of Vegetation Science* 4:223–234.

de Souza, F. M., and Batista, J. L. F. 2004. Restoration of seasonal semi-deciduous forests in Brazil: Influence of age and restoration design on forest structure. *Forest Ecology and Management* 191:185–200.

Díaz, S, Cabido, M., Zak, M., Martinéz Carretero, E., and Araníbar, J. 1999. Plant functional traits, ecosystem structure and land-use history along a climatic gradient in central-western Argentina. *Journal of Vegetation Science* 10:651–660.

Díaz, S., Hodgson, J., Thompson, G. K., Cabido, M, Comelissen, J. H. C., Jalili, A., Montserrat-Marti, G. Grime, J. P., Zarrinkamer, F., Asri, Y., Band, S. R., Basconcelo, S., Castro-Diez, P., Funes, P., Hamzehee, B., Khoshnevi, M., Harguindeguy, N., Perez-Rontome, M. C., Shirvany, F. A., Vendramini, F., Yazdani, S., Abbas-Azimi, R., Bogaard, A., Boustani, S., Charles, M., Dehghan, M., de Torres-Espuny, L., Falczuk, V., Guerrero-Campo, J., Hynd, A., Jones, G., Kowsary, E., Kazemi-Saeed, F., Maestro-Martinez, M., Romo-Diez, A., Shaw, S., Siavash, B., Villar-Salvador, P., and Zak, M. R. 2004. The plant traits that drive ecosystems: Evidence from three continents. *Journal of Vegetation Science* 15:295–304.

Dostal, P. 2005. Effect of three mound-building ant species on the formation of soil seed banks in mountain grassland. *Flora* 200:148–158.

Egler, F. E. 1954. Vegetation science concepts: I. Initial floristic composition, a factor in old-field vegetation development. *Vegetatio* 4:412–417.

Fargione, J. E., and Tilman, D. 2005. Diversity decreases invasion via both sampling and complementarity effects. *Ecology Letters* 8:604–611.

Fattorini, M., and Halle, S. 2004. The dynamic environmental filter model: How do filtering effects change in assembling communities after disturbance? In: *Assembly Rules and Restoration Ecology*. V. M. Temperton, R. J. Hobbs, T. J. Nuttle, and S. Halle (eds.). Washington, D.C.: Island Press, pp. 96–114.

Feldpausch, T. R., Rondon, M. A., Femandes, E. C. M., Riha, S. J., and Wandelli, E. 2004. Carbon and nutrient accumulation in secondary forests regenerating on pastures in central Amazonia. *Ecological Applications* 14:S164–S176.

Fike, J., and Niering, W. A. 1999. Four decades of old field vegetation development and the role of *Celastrus orbiculatus* in the northeastern United States. *Journal of Vegetation Science* 10:483–492.

Fukami, T., Bezemer, T. M., Mortimer, S. R., and van der Putten, W. H. 2005. Species divergence and trait convergence in experimental plant community assembly. *Ecology Letters* 8:1283–1290.

Fuller, R. N., and del Moral, R. 2003. The role of refugia and dispersal in primary succession on Mount St. Helens, Washington. *Journal of Vegetation Science* 14:637–644.

Garcia, D., and Obeso, J. R. 2003. Facilitation by herbivore-mediated nurse plants in a threatened tree, *Taxus baccata*: Local effects and landscape level consistency. *Ecography* 26:739–750.

Gorb, S. N., Gorb, E. V., and Punttila, P. 2000. Effects of redispersal of seeds by ants on the vegetation pattern in a deciduous forest: A case study. *Acta Oecologica* 21:293–301.

Gosling, P. 2005. Facilitation of *Urtica dioica* colonisation by *Lupinus arboreus* on a nutrient-poor mining spoil. *Plant Ecology* 178:141–148.

Gotelli, N. J. 1991. Metapopulation models-the rescue effect, the propagule rain, and the core-satellite hypothesis. *American Naturalist* 138:768–776.

Grime, J. P. 2001. *Plant Strategies, Vegetation Processes, and Ecosystem Properties.* Chichester, U.K.: Wiley.

Hawkes, C. V., and Sullivan, J. J. 2001. The impact of herbivory on plants in different resource conditions: A meta-analysis. *Ecology* 82:2045–2058.

Hector, A., Schmid, B., Beierkuhnlein, C., Caldeira, M. C., Diemer, M., Dimitrakopoulos, P. G., Finn, J. A., Freitas, H., Giller, P. S., Good, J., Harris, R., Hogberg, P., Huss-Danell, K., Joshi, J., Jumpponen, A., Korner, C., Leadley, P. W., Loreau, M., Minns, A., Mulder, C. P. H., O'Donovan, G., Otway, S. J., Pereira, J. S., Prinz, A., Read, D. J., Scherer-Lorenzen, M., Schulze, E. D., Siamantziouras, A. S. D., Spehn, E. M., Terry, A. C., Troumbis, A. Y., Woodward, F. I., Yachi, S., and Lawton, J. H. 1999. Plant diversity and productivity experiments in European grasslands. *Science* 285:1123–1127.

Henríquez, J. M., and Lusk, C. H. 2005. Facilitation of *Nothofagus antarctica* (Fagaceae) seedlings by the prostrate shrub *Empetrum rubrum* (Empetraceae) on glacial moraines in Patagonia. *Austral Ecology* 30:885–890.

HilleRisLambers, J., Harpole, W. S., Tilman, D., Knops, J., and Reich, P. B. 2004. Mechanisms responsible for the positive diversity-productivity relationship in Minnesota grasslands. *Ecology Letters* 7:661–668.

Hodkinson, I. D., Webb, N. R., and Coulson, S. J. 2002. Primary community assembly on land -the missing stages: Why are the heterotrophic organisms always there first? *Journal of Ecology* 90:569–577.

Holl, K. D., Loik, M. E., Lin, E. H. V., and Samuels, I. A. 2000. Tropical montane forest restoration in Costa Rica: Overcoming barriers to dispersal and establishment. *Restoration Ecology* 8:339–349.

Hölzel, N., and Otte, A. 2003. Restoration of a species-rich flood meadow by topsoil removal and diaspore transfer with plant material. *Applied Vegetation Science* 6:131–140.

Honnay, O., Verhaeghe, W., and Hermy, M. 2001. Plant community assembly along dendritic networks of small forest streams. *Ecology* 82:1691–1702.

Honnay, O., Verheyen, K., and Hermy, M. 2002. Permeability of ancient forest edges for weedy plant species invasion. *Forest Ecology and Management* 161:109–122.

Hooper, E. R., Legendre, P., and Condit, R. 2004. Factors affecting community composition of forest regeneration in deforested, abandoned land in Panama. *Ecology* 85:3313–3326.

Hooper, E. R., Legendre, P., and Condit, R. 2005. Barriers to forest regeneration of deforested and abandoned land in Panama. *Journal of Applied Ecology* 42:1165–1174.

Howe, H. F., and Lane, D. 2004. Vole-driven succession in experimental wet-prairie restorations. *Ecological Applications* 14:1295–1305.

Huttl, R. F., and Weber, E. 2001. Forest ecosystem development in post-mining landscapes: a case study of the Lusatian lignite district. *Naturwissenschaften* 88:322–329.

Jones, A. S., Lamont, B. B., Fairbanks, M. M. and Rafferty, C. M. 2003. Kangaroos avoid eating seedlings with or near others with volatile essential oils. *Journal of Chemical Ecology* 29:2621–2635.

Jones, C. C., and del Moral, R. 2005. Effects of microsite conditions on seedling establishment on the foreland of Coleman Glacier, Washington. *Journal of Vegetation Science* 16:293–300.

Khater, C., Martin, A., and Maillet, J. 2003. Spontaneous vegetation dynamics and restoration prospects for limestone quarries in Lebanon. *Applied Vegetation Science* 6:199–204.

Knevel, I. C., Bekker, R. M., Kleyer, M., and Bakker, J. P. 2003. Life-history traits of the Northwest European flora: A data-base (LEDA). *Journal of Vegetation Science* 14:611–614.

Koch, J. M., Richardson, J., and Lamont, B. B. 2004. Grazing by kangaroos limit the establishment of the grass trees *Xanthorrhoea gracilis* and *X. preissii* in restored bauxite mines in Eucalypt forests of Southwestern Australia. *Restoration Ecology* 12:297–305.

Kochy, M., and Wilson, S. D. 2005. Variation in nitrogen deposition and available soil nitrogen in a forest-grassland ecotone in Canada. *Landscape Ecology* 20:191–202.

Lane, D. R., and Bassiri Rad, H. 2005. Diminishing spatial heterogeneity in soil organic matter across a prairie restoration chronosequence. *Restoration Ecology* 13:403–412.

Lockwood, J. L., and Pimm, S. L. 1999. When does restoration succeed? In: *Ecological Assembly: Advances, Perspectives, Retreats*. E. Weiher and P. Keddy (eds.). Cambridge: Cambridge University Press, pp. 363–392.

Margules, C. R., and Pressey, R. L. 2000. Systematic conservation planning. *Nature* 405:243–253.

Martínez-Garza, C., and Howe, H. F. 2003. Restoring tropical diversity: Beating the time tax on species loss. *Journal of Applied Ecology* 40:423–429.

McCune, B., and Allen, T. F. H. 1985. Will similar forests develop on similar sites? *Canadian Journal of Botany* 63:367–376.

McEuen, A. B., and Curran, L. M. 2004. Seed dispersal and recruitment limitation across spatial scales in temperate forest fragments. *Ecology* 85:507–518.

Morgantini, L. L., and Kansas, J. L. 2003. Differentiating mature and old-growth forests in the upper foothills and subalpine subregions of west-central Alberta. *Forestry Chronicle* 79:602–612.

Murakami, K., Maenaka, H., and Morimoto, Y. 2005. Factors influencing species diversity of ferns and fern allies in fragmented forest patches in the Kyoto city area. *Landscape and Urban Planning* 70:221–229.

Niering, W. A., and Egler, F. E. 1955. A shrub community of *Viburnum lentago*, stable for twenty-five years. *Ecology* 36:356–360.

Ogden, J. A. E., and Rejmanek, M. 2005. Recovery of native plant communities after the control of a dominant invasive plant species, *Foeniculum vulgare*: Implications for management. *Biological Conservation* 125:562–568.

Palmer, J. P., and Chadwick, M. J. 1985. Factors affecting the accumulation of nitrogen in colliery spoil. *Journal of Applied Ecology* 22:249–257.

Piessens, K., Honnay, O., Nackaerts, K., and Hermy, M. 2004. Plant species richness and composition of heathland relics in north-western Belgium: Evidence for a rescue-effect? *Journal of Biogeography* 31:1683–1692.

Polley, H. W., Derner, J. D., and Wilsey, B. J. 2005. Patterns of plant species diversity in remnant and restored tallgrass prairies. *Restoration Ecology* 13:480–487.

Prach, K., Bartha, S., Joyce, C. B., Pyšek, P., van Diggelen, P., and Wiegleb, G. 2001a. The role of spontaneous vegetation in ecosystem restoration: A perspective. *Applied Vegetation Science* 4:111–114.

Prach, K., Pyšek, P., and Bastl, M. 2001b. Spontaneous vegetation succession in human-disturbed habitats: A pattern across seres. *Applied Vegetation Science* 4:83–88.

Pyšek, P., Chocholouskova, Z., Pyšek, A., Jarosik, V., Chytry, M., and Tichy, L. 2005. Trends in species diversity and composition of urban vegetation over three decades. *Journal of Vegetation Science* 15:781–788.

Ramsey, D. S. L., and Wilson, J. C. 1997. The impact of grazing by macropods on coastal foredune vegetation in southeast Queensland. *Australian Journal of Ecology* 22:288–297.

Rayfield, R., Anand, M., and Laurence, S. 2005. Assessing simple versus complex restoration strategies for industrially disturbed forests. *Restoration Ecology* 13:639–650.

Reid, N., and Naeth, M. A. 2005. Establishment of a vegetation cover on tundra kimberlite mine tailings: 2. A field study. *Restoration Ecology* 13:602–609.

Rodwell, J. (ed.). 1991–2000. *British Plant Communities*. Vol. 1–5. Cambridge: Cambridge University Press.

Roscher, C., Temperton, V. M., Scherer-Lorenzen, M., Schmitz, M., Schumaher, J., Schmid, B., Buchmann, N., Weisser, W. W., and Schulze, E. D. 2005. Over yielding in experimental grassland communities—irrespective of species pool or spatial scale. *Ecology Letters* 8:576–577.

Rosenfeld, J. S. 2002. Functional redundancy in ecology and conservation. *Oikos* 98:156–162.

Ross, K. A., Fox, B. J., and Fox, M. D. 2002. Changes to plant species richness in forest fragments: Fragment age, disturbance and fire history may be as important as area. *Journal of Biogeography* 29:749–765.

Schaminée J. H. J. (ed.). 1995–1999. *De Vegetatie van Nederland*. Vol. 1–5. Uppsala: Opulus Press.

Schwartz, M. W., Brigham, C. A., Hoeksema, J. D., Lyons, K. G., Mills, M. H., and van Mantgem, P. J. 2000. Linking biodiversity to ecosystem functions: Implications for conservation ecology. *Oecologia* 122:297–305.

Sekura, L. S., Mal, T. K., and Dvorak, D. F. 2005. A long-term study of seedling regeneration for an oak forest restoration in Cleveland Metroparks Brecksville Reservation, Ohio. *Biodiversity and Conservation* 14:2397–2418.

Shiels, A. B., and Walker, L. R. 2003. Bird perches increase forest seeds on Puerto Rican landslides. *Restoration Ecology* 11:457–465.

Slocum, M. G., and Horvitz, C. C. 2000. Seed arrival under different genera of trees in a neotropical pasture. *Plant Ecology* 149:51–62.

Smith, M. D., and Knapp, A. K. 2003. Dominant species maintain ecosystem function with non-random species loss. *Ecology Letters* 6:509–517.

Snyman, H. A. 2003. Revegetation of bare patches in a semi-arid rangeland of South Africa: An evaluation of various techniques. *Journal of Arid Environments* 55:417–432.

Spehn, E. M., Hector, A., Joshi J., Scherer-Lorenzen, M., Schmid, B., Bazeley-White, E., Beierkuhnlein, C., Caldeira, M. C., Diemer, M., Dimitrakopoulos, P. G., Finn, J. A., Freitas, H., Giller, P. S., Good, J., Harris, R., Hogberg, P., Huss-Danell, K., Jumpponen, A., Koricheva, J., Leadley, P. W., Loreau, M., Minns, A., Mulder, C. P. H., O'Donovan, G., Otway, S. J., Palmborg, C., Pereira, J. S., Pfisterer, A. B., Prinz,

A., Read, D. J., Schulze, E. D., Siamantziouras, A. S. D., Terry, A. C., Troumbis, A. Y., Woodward, F. I., Yachi, S., and Lawton, J. H. 2005. Ecosystem effects of biodiversity manipulations in European grasslands. *Ecological Monographs* 75:37–63.

Suding, K. N., Collins, S. L., Gouch, L., Clark, C., Cleland, E. E., Gross, K. L., Milchunas, D. G., and Pennings, S. 2005. Functional- and abundance-based mechanisms explain diversity loss due to N fertilization. *Proceedings of the National Academy of Sciences* 102:4387–4392.

Svenning, J. C., and Wright, S. J. 2005. Seed limitation in a Panamanian forest. *Journal of Ecology* 93:853–862.

Symstad, A. J., and Tilman, D. 2001. Diversity loss, recruitment limitation, and ecosystem functioning: Lessons learned from a removal experiment. *Oikos* 92:424–435.

Symstad, A. J., Chapin, F. S., Wall, D. H., Gross, K. L., Huenneke, L. F., Mittelbach, G. G., Peters, D. P. C., and Tilman, D. 2003. Long-term and large-scale perspectives on the relationship between biodiversity and ecosystem functioning. *BioScience* 53:89–98.

Taverna, K., Peet, R. K., and Phillips, L. C. 2005. Long-term change in ground-layer vegetation of deciduous forests of the North Carolina Piedmont, USA. *Journal of Ecology* 93:202–213.

Temperton, V. M., Hobbs, R. J., Nuttle, T., and Halle, S. (eds.). 2004. *Assembly Rules and Restoration Ecology*. Washington, D.C.: Island Press.

Temperton, V. M., and Zirr, K. 2004. Order of arrival and availability of safe sites. In: *Assembly Rules and Restoration Ecology*. V. M. Temperton, R. J. Hobbs, T. Nuttle, and S. Halle (eds.). Washington, D.C.: Island Press, pp. 285–304.

Toh, I., Gillespie, M., and Lamb, D. 1999. The role of isolated trees in facilitating tree seedling recruitment at a degraded sub-tropical rainforest site. *Restoration Ecology* 7:288–297.

Tsuyuzaki, S., Titus, J. T., and del Moral, R. 1997. Seedling establishment patterns on the Pumice Plain, Mount St. Helens, Washington. *Journal of Vegetation Science* 8:727–734.

Turner, M. G., Baker, W. L., Peterson, C. J., and Peet, R. K. 1998. Factors influencing succession: Lessons from large, infrequent natural disturbances. *Ecosystems* 1:511–523.

Vail, S. G. 1992. Selection for over-compensatory plant responses to herbivory: A mechanism for the evolution of plant-herbivore mutualism. *American Naturalist* 139:1–8.

Van Andel, J., and Aronson, J. 2006. *Restoration Ecology – The New Frontier*. Oxford, U. K.: Blackwell.

van der Heijden, M. G. A., Klironomos, J. N., Ursic, M., Mountoglis, P., Streitwolf-Engel, R., Boller, T., Wiemken, A., and Sanders, I. R. 1998. Mycorrhizal fungal diversity determines plant biodiversity, ecosystem variability and productivity. *Nature* 396:69–72.

Verheyen, K., and Hermy, M. 2004. Recruitment and growth of herb-layer species with different colonizing capacities in ancient and recent forests. *Journal of Vegetation Science* 15:125–134.

Wagner, M. 2004. The roles of seed dispersal ability and seedling salt tolerance in community assembly of a severely degraded site. In: *Assembly Rules and Restoration Ecology*. V. M. Temperton, R. J. Hobbs, T. Nuttle, and S. Halle (eds.). Washington, D.C.: Island Press, pp. 266–284.

Walker, L. R. 1993. Nitrogen fixers and species replacement in primary succession. In: *Primary Succession on Land*. J. Miles and D. W. H. Walton (eds.). Oxford, U.K.: Blackwell, pp. 249–272.

Walker, L. R., and del Moral, R. 2003. *Primary Succession and Ecosystem Rehabilitation*. Cambridge, U.K.: Cambridge University Press.

Walker, L. R., Bellingham, P. J., and Peltzer, D. A. 2006. Plant characteristics are poor predictors of microsite colonization during the first two years of primary succession. *Journal of Vegetation Science* 17:397–406.

Weiher, E., and Keddy, P. 1995. The assembly of experimental wetland plant communities. *Oikos* 73:323–335.

White, E., Tucker, N., Meyers, N., and Wilson, J. 2004. Seed dispersal to revegetated isolated rainforest patches in North Queensland. *Forest Ecology and Management* 192:409–426.

Wolters, M., Garbutt, A., and Bakker, J. P. 2005. Salt-marsh restoration: Evaluating the success of de-embankments in north-west Europe. *Biological Conservation* 123:249–268.

Young, T. P., Person, D. A., and Clary, J. J. 2005. The ecology of restoration: Historical links, emerging issues and unexplored realms. *Ecology Letters* 8:662–673.

Zanini, L., and Ganade, G. 2005. Restoration of *Araucaria* forest: The role of perches, pioneer vegetation, and soil fertility. *Restoration Ecology* 13:507–514.

Zavaleta, E. S., Shaw, M. R., Chiariello, N. R., Mooney, H. A., and Field, C. B. 2003. Additive effects of simulated climate changes, elevated CO_2 and nitrogen deposition on grassland diversity. *Proceedings of the National Academy of Sciences* 100:7650–7654.

Aboveground–Belowground Linkages, Ecosystem Development, and Ecosystem Restoration

David A. Wardle and Duane A. Peltzer

Key Points

1. All ecosystems consist of aboveground and belowground components that interact with each other to drive community and ecosystem properties. The feedbacks between these two components are therefore potentially useful for understanding the principles of succession and restoration.
2. We provide three case studies in which understanding aboveground–belowground feedbacks are relevant for succession and restoration. These involve human induced changes in densities of browsing herbivores with particular reference to deer in New Zealand forests; the impacts of fire and fire suppression with particular reference to boreal forests in northern Sweden; and the belowground impacts of invasive nonnative plants and their feedbacks aboveground.
3. Finally, we explore the utility of the aboveground–belowground model as an approach that can help us understand successional processes and that can be incorporated into restoration efforts. In doing this we also propose profitable areas of future research.

3.1 Introduction

All terrestrial ecosystems consist of explicit aboveground and belowground biotic components. Although these have traditionally been considered in isolation from one another, there has been increasing recognition over the past decade or so that these components interact with each other to drive processes at both the community and ecosystem levels of resolution (e.g., van der Putten *et al.* 2001, Wardle *et al.* 2004a, Bardgett 2005). Plants (primary producers) provide the input of carbon required by the decomposer community, while the decomposers in turn break down organic matter and thus regulate the supply of available nutrients for the plants. Further, aboveground herbivores, that biota associated with live roots (pathogens, root herbivores, and mutualists), and their predators exert important effects on feedbacks between the aboveground and belowground subsystems. Over the past two decades there has been increasing recognition that biotic factors are fundamental determinants of the functioning of terrestrial ecosystems (Vitousek *et al.* 1987, Lawton 1994, Chapin *et al.* 1997), and feedbacks between the aboveground and belowground components are arguably among the most important of these biotic drivers (Wardle 2002).

Development of a thorough understanding of either primary or secondary ecological succession, and ecosystem restoration, requires specific consideration of both the aboveground and belowground subsystems, as well as the nature of feedbacks between them. Although primary successional development on newly created surfaces, or secondary succession following significant disturbance events, has usually been studied only with specific reference to the plant community and the availability of major soil nutrients (Bradshaw and Chadwick 1980), the reality is that the aboveground and belowground communities develop in close concert with each other over successional time. Further, plant species replacement (such as occurs both during primary and secondary succession) has major effects on soil communities and the ecological processes that they control (Wardle *et al.* 1999, Porazinska *et al.* 2003, Belnap *et al.* 2005). Changes in soil communities in turn influence the direction and speed of both primary and secondary vegetation succession as well as ecosystem productivity (van der Putten *et al.* 1993, De Deyn *et al.* 2003). Knowledge about feedbacks between aboveground and belowground biota is also crucial to developing a better understanding of the principles of ecosystem restoration, because facilitating the recovery of ecosystems requires recognition of the role of these feedbacks in driving community- and ecosystem-level properties and processes.

In this chapter, we will start by discussing vegetation succession and ecosystem development in the context of a combined aboveground–belowground approach. We will then present three case studies in which principles of succession studied through a combined aboveground–belowground approach are relevant to the goals of ecosystem restoration: (1) the consequences of human-induced changes in densities of browsing mammals, with particular reference to deer in New Zealand rain forests; (2) the ecological impacts of fire in the long-term, with particular reference to boreal forests of northern Sweden; and (3) the belowground impacts of invasive, nonnative plant species and their feedbacks aboveground. These examples will be used to emphasize the importance of combined aboveground–belowground approaches to understanding vegetation succession, the ecological role of disturbance, and the restoration of ecological interactions and processes.

3.2 Successional Development and Aboveground–Belowground Linkages

An important component of ecosystem development and succession is the changes that occur in the attributes of the dominant vegetation. As primary succession proceeds, there is a general shift in dominant plant species from those that are of small stature, often herbaceous, short-lived, have a high reproductive output, and produce litter with high quality with those that are increasingly larger, woody, long lived, more conservative in retaining nutrients, and produce foliage and litter of poorer quality (Grime 1979, Walker and Chapin 1987, Wardle 2002). Similar trends also occur in secondary succession, even though the starting point may be later in ecosystem development because of legacy effects. These changes in vegetation composition have important consequences for resource input to the soil. For example, in the first few years of primary succession, nitrogen input often increases rapidly as a result of colonization by plant species that are capable of forming symbiotic relationships with bacteria that fix atmospheric N_2 (Bradshaw and Chadwick 1980). This is apparent, for example, through colonization of *Lupinus* spp. on Mt. St. Helens (Morris and Wood 1989), *Alnus* spp. on fresh floodplains (Luken and Fonda 1983), and *Carmichaelia* on newly created gravel outwashes (Bellingham *et al.* 2001). Concomitant with this are rapid increases in net primary productivity (NPP) and net accumulation of organic matter in the soil (Schlesinger *et al.* 1998). This accumulation is made possible through the steady accumulation of microbial residues and humified materials, which facilitate soil moisture availability and nutrient cycling.

Over the past few years, a growing number of studies have investigated how plant species and community composition affect the community composition of the soil biota (e.g., Wardle *et al.* 1999, Porazinska *et al.* 2003). There is also recognition that traits of dominant plant species have important indirect consequences for the belowground subsystem (Wardle *et al.* 1998, Eviner and Chapin 2003), including the densities and community composition of soil organisms (De Deyn *et al.* 2004, Vitecroft *et al.* 2005). These impacts are especially relevant for understanding belowground changes that occur during either primary or secondary vegetation succession, as this literature suggests that changes in the functional composition of vegetation during succession should be matched belowground. As such, microbial communities during ecosystem development in both herbaceous and woody systems show shifts from bacterial to fungal domination (e.g., Ohtonen *et al.* 1999), an attribute that is frequently associated with greater conservation of nutrients (Coleman *et al.* 1983). Other changes in soil communities that have been identified during succession include shifts in nematode taxa from those known to be r-selected to those that are K-selected (Wasilewska 1994), shifts from domination by arbuscular mycorrhizal fungi to ectomycorrhizal fungi (Dighton and Mason 1985), increases in the length of soil food chains (Verhoeven 2002), and enrichment of diversity within trophic groups (Sigler and Zeyer 2002, Dunger *et al.* 2004). There is also evidence that during vegetation succession, changes in the community structure of at least some belowground groups may be deterministic and consistent across succession (Hodkinson *et al.* 2004). It is, therefore, apparent that changes in the quantity and quality of resources present in soils during the course of vegetation succession will have important, and often predictable, consequences for belowground communities.

Just as the aboveground biota influences the belowground biota, so the belowground biota influences what we see aboveground. Soil organisms affect plant communities both through a direct pathway and an indirect pathway (Wardle *et al.* 2004a, van der Putten 2005) (Fig. 3.1). The direct pathway involves those organisms that affect plants through being intimately associated with plant roots (e.g., mycorrhizal fungi, root herbivores, root pathogens) while the indirect pathway involves decomposer biota that indirectly affect plant growth through mineralizing or immobilizing plant-available nutrients. Although the literature is replete with examples that show how both pathways may affect plant growth, few studies have investigated how belowground community composition affects plant communities. However, soil pathogens affect successional trajectories in

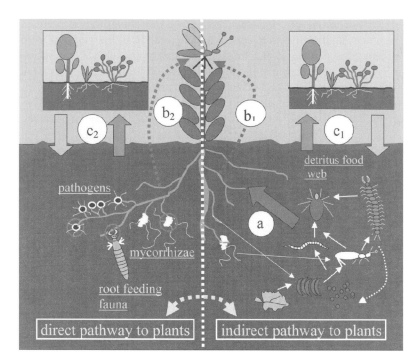

Figure 3.1 Aboveground communities as affected by both direct and indirect consequence of soil food web organisms. (Right): Feeding activities in the detritus food web
(slender solid arrows) stimulate nutrient turnover, plant nutrient acquisition (a), and plant
performance, and thereby indirectly influence aboveground herbivores (broken arrow)
(b_1). (Left): Soil biota exert direct effects on plants by feeding upon roots and forming
antagonistic or mutualistic relationships with their host plants. Such direct interactions
with plants influence not just the performance of the host plants themselves but also
that of the herbivores (b_2) and potentially their predators. Further, the soil food web
can control the successional development of plant communities both directly (c_2) and
indirectly (c_1), and these plant community changes can in turn influence soil biota. Reproduced from Wardle *et al.* (2004a) with permission from the American Association
of the Advancement of Science. Drawing by Heikki Setälä.

foredune communities (van der Putten *et al.* 1993), and root herbivores affect the relative abundance of different plant functional groups (Brown and
Gange 1990). Further, the structure of arbuscular mycorrhizal fungal communities serves as a driver of plant community structure (Van der Heijden *et al.*
1998). In a mesocosm study, De Deyn *et al.* (2003) showed that soil invertebrate
fauna suppress early successional grassland plant species and thus promote the
rate of secondary successional change in vegetation. Less is understood about
how the community structure of decomposer microbes and fauna affects either
primary or secondary vegetation succession. However, given that decomposer
community structure influences nutrient mineralization and therefore plant nutrient acquisition and growth (Laakso and Setälä 1999, Liiri *et al.* 2002), it is
probable that the plant community composition is also responsive, with early
successional plants (that are most dependent on mineral nutrients) benefiting
most from those soil organisms that maximize soil mineralization.

Succession is characterized by shifts not only at the community level but also
at the ecosystem level (Odum 1969). Aboveground–belowground feedbacks

contribute to maintaining ecosystem-level properties through influencing key ecosystem functions such as NPP, decomposition, and nutrient flux. As succession proceeds, NPP increases sharply, but there is an increasing dominance of those plant species that produce poorer litter quality (i.e., higher carbon to nutrient ratios, greater levels of defense compounds), and thus retard decomposition and mineralization processes. This is associated with an increasing importance of those soil organisms that are associated with a tighter and more conservative cycling of nutrients in the ecosystem, for example, domination by fungi over bacteria, and ectomycorrhizal fungi over arbuscular mycorrhizal fungi (Coleman *et al.* 1983). There is also a lower turnover of microbial tissues, and a lower proportion of the microbial biomass is utilized by consumers (Wardle 2002). The net result is lower levels of available mineral nutrients in the soil, greater accumulation of soil organic matter, lower NPP, and a slower turnover of C and nutrients both aboveground and belowground through succession.

When ecosystems are left for significant periods without catastrophic disturbance (i.e., thousands to tens of thousands of years), both primary and secondary seres can proceed to a state of ecosystem retrogression (see Chapter 4). This is often driven by a reduction of available phosphorus (P) over time (Walker and Syers 1976, Vitousek 2004), and appears to occur in broadly similar ways across vastly different ecosystems and successional types. In a study of six long-term seres that ranged from the boreal zone to the tropics and for which a retrogressive phase was present, substantial reductions in plant biomass over time were matched with increasing limitation of P relative to N, and reduced performance of the decomposer subsystem (Wardle *et al.* 2004b). Often this was also matched by shifts in microbial community structure, and increasing domination of bacteria relative to fungi in the oldest systems. Although aboveground and belowground biota (and the processes that they drive) may show similar patterns of decline during retrogression, much remains unknown about the involvement of aboveground–belowground feedbacks in retrogressive processes.

Ecosystem restoration is focused on attempting to reverse human-induced damage to changes in community- and ecosystem-level properties and processes, and therefore represents an obvious application of the principles of succession (including both primary and secondary succession). The aboveground–belowground approach has proved its worth in several studies that have aimed to understand how human activities affect communities and ecosystems, for example, through land use change (Compton and Boone 2000), atmospheric CO_2 enrichment (Díaz *et al.* 1993), N deposition (Aerts and Berendse 1988), global warming (Seastedt 2000), biological invasions (Ehrenfeld 2003), and biodiversity loss (Wardle *et al.* 1999). Such studies enable us to predict the state that aboveground and belowground properties of ecosystems should converge to over time when the human-induced impact in question is minimized or removed. This is, in turn, useful in allowing us to better understand the successional trajectory that might be expected or encouraged, both at the community and ecosystem level, when restoration efforts to allow recovery from human-induced impacts are implemented. Further, by comparing our understanding of ecological interactions that occur along successional gradients (both primary and secondary), we can better understand the starting and desired end points of any restoration activity, and then try to mimic successional patterns from that starting point.

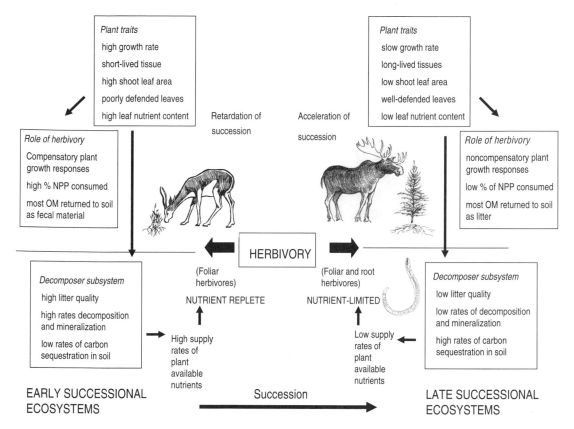

Figure 3.2 Mechanistic basis of how herbivores affect the decomposer subsystem at a plant community level, through altering successional trajectories. Reproduced from Bardgett and Wardle (2003) with permission from the Ecological Society of America.

3.3 Browsing Mammals and New Zealand Rainforests

Foliar herbivory is an important driver of many ecosystems, and depending on the ecosystem considered, between 1 and 50% of NPP is consumed by herbivores (McNaughton *et al.* 1989). The intensity of herbivory is influenced by succession, with earlier successional systems often being subjected to a greater intensity of herbivory than later successional systems. Further, herbivores are important in influencing the direction and rate of vegetation succession; in fertile systems, herbivores retard succession while in infertile systems they promote succession (Bardgett and Wardle 2003) (Fig. 3.2).

Aboveground herbivores influence not just plant growth and plant communities, but also indirectly affect decomposers and decomposer processes across a range of temporal and spatial scales (Bardgett *et al.* 1998, Wardle and Bardgett 2004). In the short-term, herbivory can induce significant flow of C from the plant to the rhizosphere microflora, creating an aboveground feedback through enhancing N availability for the plants (Hamilton and Frank 2001). In the longer

term, herbivores can stimulate decomposers by promoting compensatory plant growth (and hence NPP), returning organic matter as labile fecal material, enhancing foliar nutrient concentrations and impairing succession, thus preventing domination by later successional species that produce more recalcitrant litter (McNaughton 1985, Augustine and McNaughton 1998). Herbivores can also depress decomposers through reducing NPP of plant species by tissue removal, inducing production of defense compounds, and promoting succession (thus encouraging domination of plant species that produce more recalcitrant litter) (Pastor *et al.* 1993). At the landscape scale the most important belowground effect of herbivores is usually through alteration of vegetation successional pathways. The effects of herbivores on vegetation succession (Bardgett and Wardle 2003, Sankaran and Augustine 2004; Fig. 3.2) in turn alter the densities of decomposer organisms and rates of processes that they regulate.

Human-induced changes in densities of browsing mammals have important consequences for both the aboveground and belowground subsystems worldwide; these include the introduction of mammals to new regions (where they function as invaders), the promotion of conditions that allow native mammals to reach unnaturally high densities, and the reduction of natural mammal populations, sometimes to extinction (reviewed by Wardle and Bardgett 2004). For each of these mechanisms, several examples exist of how human-induced shifts in densities of browsing mammals have affected vegetation succession and consequently the decomposer biota. This is indicative of important herbivore-induced effects on feedbacks between the aboveground and belowground subsystems, and collectively the impacts of these herbivores may represent an important, though often unrecognized, component of global change (Zimov *et al.* 1995, Wardle and Bardgett 2004, Burney and Flannery 2005). Restoration activities would be aimed at returning mammal densities to the levels at which they would occur in the absence of human activity (either by increasing or reducing their densities), and thus encouraging the ecosystem to more closely resemble its prehuman condition. It is apparent that alteration of mammal densities and therefore intensity of herbivory will in turn alter the successional trajectory of the vegetation.

New Zealand rainforests provide excellent opportunities to study the consequences of human-induced alterations of herbivore densities on aboveground–belowground feedbacks. Several species of large browsing mammals were introduced to New Zealand between the 1770s and 1920s, the most pervasive of which are the European red deer and feral goats. Before human settlement, browsing mammals did not exist in New Zealand, making this system ideal for investigating the impacts of introducing a whole functional group of animals to an environment where they were previously absent. Further, in the 1950s to 1980s the former New Zealand Forest Service established several hundred fenced exclosure plots (typically 20m × 20m) throughout New Zealand's forests to assess browsing mammal impacts on vegetation; a subset of these are still effective and they provide real opportunities for studying ecological impacts of browsing mammals over the order of decades. Through exclosure studies, it has been demonstrated that deer have important and consistent effects on vegetation composition, through consuming and removing fast growing, palatable, broad-leaved species, and encouraging their replacement by unpalatable monocotyledonous species, ferns, and small-leaved species (Conway 1949, Wallis and James 1972). This is consistent with the model of herbivores promoting succession toward domination by plant species that are slow growing and well defended (Fig. 3.2). These results contribute to understanding restoration in a successional context, because they give insights as to the types of successional changes that might be expected in vegetation when invasive vertebrate species are removed.

These effects of deer on vegetation in turn affect the quality of organic matter entering the decomposer subsystem. Analysis of feeding preferences by deer in New Zealand forests provides consistent evidence that they prefer plant species with low levels of foliar lignin and fiber (Forsyth *et al.* 2002 and 2005). Further, Wardle *et al.* (2002) found, using 30 exclosure fenced plots located throughout New Zealand, that those plant species consistently reduced by deer produced litter which had lower concentrations of lignin, condensed tannins, and polyphenols than those species promoted by deer. Litter from those species that the deer preferred also decomposed more rapidly, and promoted the decomposition of litter from other species when placed in mixtures with them. This indicates that deer promote replacement of plant species with high litter quality (and therefore litter likely to favor decomposers) with litter of poor quality. Ecosystem restoration by removal of deer could conceivably reverse this successional pathway, by maintaining an increased density of vegetation that is favorable for microbes. However, it is important to note that this is only one mechanism by which browsers may affect decomposers (Bardgett and Wardle 2003), and other mechanisms that work in different directions may also be involved.

Despite the consistent nature of trends observed in the aboveground community for these 30 exclosures, the response of much of the belowground biota was far less predictable (Wardle *et al.* 2001). Most of the main groups of small-bodied soil organisms such as the microflora and microfauna (e.g., nematodes, copepods, rotifers) showed variable responses to the presence of deer; both positive and negative significant effects occurred depending upon location and forest type. A similar pattern was found for soil properties influenced by decomposer organisms, such as sequestration of C and N in the soil, and rates of soil C mineralization. However, browsing mammals did have consistently adverse effects on densities of large-bodied soil organisms, such as the main groups of mesofauna (mites and springtails) and macrofauna (e.g., spiders, opilionids, gastropods, millipedes, beetles). The varied effect of deer on small-bodied organisms is explicable in terms of browsers affecting them through a variety of mechanisms, some positive and some negative, with different mechanisms dominating in different contexts (Bardgett and Wardle 2003, Sankaran and Augustine 2004). The consistently negative effect on large-bodied soil organisms is most likely due to physical disturbances created by deer such as trampling. Small-bodied soil organisms are more likely than large-bodied ones to be protected against such disturbances. Although these exclosure plots have only been in operation for a few decades at most (i.e., since the 1960s), they nevertheless point to strong belowground effects of deer on decomposer biota, and over time this would be expected to influence supply rates of plant-available nutrients from the soil, thus creating important feedback effects aboveground. These findings are relevant to restoration, because they show that alterations of successional pathways through the removal of invasive deer species should simultaneously exert important effects on both the aboveground and belowground subsystems, and therefore the feedbacks that exist between the two components over the longer term.

As discussed, restoration of New Zealand forests, and promotion of successional changes in these forests to a less human-modified condition, would in the first instance require the removal of browsing mammals. Studies such as those described above enable us to predict, at both the community- and ecosystem-levels, the likely consequences of removing these mammals over the order of a few decades, and therefore the extent to which these forests can be restored. However, if restoration goals are to be focused on restoring these forests to their "natural" prehuman state, then the situation becomes more complex. This is because, prior to human settlement ca. 1000 years ago, New Zealand was dominated by moas—a guild of large browsing native birds that were hunted to extinction a few hundred years ago and for which no contemporary substitutes exist. This effectively makes the goals of ecosystem restoration (i.e., reversion of these forests to a prehuman state) unattainable. Our knowledge of what effects moas had in these forests relative to those currently exerted by browsing mammals is far from clear (Atkinson and Greenwood 1989, McGlone and Clarkson 1993), although the effects of moas were probably less than the current impacts of introduced mammals (McGlone and Clarkson 1993). Further, deer probably exert greater soil disturbance per unit body mass (and therefore have more adverse effects on litter dwelling invertebrates) than did moas, because of the relative shapes of their feet (Duncan and Holdaway 1989). In any case, it is apparent that deer and goats can exert important effects in natural forests at both the community- and ecosystem-levels, and both aboveground and belowground, through altering successional pathways in the ecosystem. Although goals to strictly restore these forests to prehuman conditions are unattainable because moas are extinct, it is apparent from the above example that significant restorative benefits to these forests are likely to result from targeted reductions in the densities of introduced mammals.

3.4 Fire Regime and Swedish Boreal Forests

Wildfire is the primary natural disturbance regime in boreal forests worldwide (Bonan and Shugart 1989), including those in Scandinavia (Niklasson and Granström 2000). Fire arrests forest successional development and prevents the system from entering long-term retrogressive phases by enabling greater availability of nutrients to rejuvenate the system (Zackrisson *et al.* 1996, De Luca *et al.* 2002a). Understanding the ecological influence of fire is therefore critical for understanding secondary succession in a wide range of ecosystems globally. Over the past 200 years, human activities have increased in the boreal forest zone of Scandinavia, and interrupted the natural fire cycle through fire suppression. Prolonged suppression of wildfire may have important effects on global carbon storage patterns, through promoting greater sequestration of carbon in the ecosystem. This may help to partly explain the so-called "missing carbon" sink, or that carbon that is evolved as CO_2 by fossil fuel burning but remains unaccounted for in global carbon budgets (Schimel 1995, Hurtt *et al.* 2002). Understanding these kinds of effects are highly relevant to restoration, because restoration of natural fire regimes in these forests, and the consequences of this for ecosystem succession, are likely to be very important as determinants of whether they act as net sources or sinks of carbon not just locally but also globally.

Table 3.1 Retrogressive successional trends that occur in lake islands in northern Sweden as a result of prolonged absence of wildfire. Sources of information are Wardle *et al.* (1997, 2003, 2004b) and Wardle and Zackrisson (2005).

Response variable	Trend
Aboveground:	
Tree vegetation	Shift from domination by *Pinus sylvestris* to *Betula pubescens* to *Picea abies*
Understory vegetation	Shift from domination by *Vaccinium myrtillus* to *Vaccinium vitis-idaea* to *Empetrum hermaphroditum*
Tree and understory biomass	Continual decline
NPP	Continual decline
Light interception by vegetation	Continual decline
Vascular plant diversity	Strong increase
Intensity of effects of plant species removals on ecosystem properties	Continual decline
Moss biomass	Slight increase
N fixation by mosses	Strong increase
Belowground:	
Polyphenol concentrations in soil	Continual increase
Decomposer microbial biomass	Continual decline
Litter decomposition rate	Continual decline
Soil carbon sequestration	Strong increase
N mineralization rate	Continual decline
Mineral N concentration	Continual decline
Ratio of mineral N to organic N	Continual decline
Ratio of soil N to P	Continual increase

The ecological impacts of long-term suppression or absence of wildfire have been investigated over the past decade through a "natural experiment" involving forested islands in Lake Uddjaure and Lake Hornavan in northern Sweden (Wardle *et al.* 1997, 2003, and 2004b, Wardle and Zackrisson 2005) (Table 3.1) (see Fig. 3.3). This system allows significant replication of discrete independent ecosystems (each island effectively operates as a separate system) at ecologically meaningful spatial scales. The main extrinsic driver that varies across these islands is wildfire caused by lightning strike; large islands burn more often than smaller ones simply because they have a larger area to be intercepted by lightning. Thus, some of the largest islands have burned in the past 100 years while some of the smallest have not burned for 5000 years. These islands therefore represent a secondary successional gradient induced by fire history. This variation across islands in fire history impacts vegetation composition. Thus, the largest (most frequently burned) islands are dominated by relatively rapid-growing early successional plant species such as *Pinus sylvestris* and *Vaccinium myrtillus*, the middle sized islands are dominated by *Betula pubescens* and *Vaccinium vitis-idaea*, and the smallest islands are dominated by slow-growing, late successional species such as *Picea abies* and *Empetrum hermaphroditum* (Wardle *et al.* 1997). *Picea* and *Empetrum* are well known to contain high levels of secondary metabolites such as polyphenols in their tissues (Nilsson and Wardle 2005), so that with increasing time since fire there is a shift from plants that invest their C in growth to those that invest C in defense. Concomitant with

Figure 3.3 A forested island in Lake Hornavan in northern Sweden.

this is a marked reduction in both plant standing biomass and NPP (Wardle *et al.* 2003). The island system thus represents a retrogressive succession with prolonged absence of disturbance leading to aboveground decline in biomass and productivity (Wardle *et al.* 2004b). Restoration of a natural fire regime would reverse this succession and thus promote biomass and productivity of these forests.

The shifts in vegetation composition aboveground have important corresponding effects belowground. For example, soils on the small islands contain higher concentrations of polyphenols than those on the large islands, presumably because of domination by *Picea* and *Empetrum* (Wardle *et al.* 1997). The quality of litter entering soils on the small islands is also inferior (Wardle *et al.* 2003). Although N inputs from N_2 fixation by cyanobacteria associated with bryophytes (the main agent of N input to these forests; De Luca *et al.* 2002b) is much greater on the small islands (Lagerström, Wardle, Nilsson, and Zackrisson, unpublished), and soils on small islands have a higher N concentration (Wardle *et al.* 1997), this N is of reduced biological availability largely because it is tightly bound in polyphenolic complexes. The net consequence of this is reduced biomass and activity of decomposer microbes, lower rates of decomposition and N mineralization, and lower concentrations of available forms of N on the small islands. Concomitant with this is reduced availability of P, a characteristic of retrogressive successions that span thousands of years (Wardle *et al.* 2004b). The reduction in soil biological activity and nutrient availability creates aboveground feedbacks, ultimately reducing plant nutrient acquisition, standing plant biomass, and NPP (Wardle *et al.* 1997 and 2003). Restoration of the natural fire regime in these forests to reverse or arrest this retrogressive succession would in turn promote soil biological activity and nutrient supply from the soil, and thus enhance NPP. This emphasizes the importance of understanding feedbacks between the aboveground and belowground subsystems when attempting to predict the ecological consequences of ecosystem restoration.

Studies on these islands have also shown that the importance of species effects on ecosystem properties changes during successional time. An ongoing experiment set up in 1996 (first 7 years reported by Wardle and Zackrisson 2005) involved experimental removals of various combinations of understory plant species on each of 30 islands across the sequence. This work revealed that two of the dominant understory species (*V. myrtillus* and *V. vitis-idaea*) drove belowground processes on the large islands but not the small ones. Removals of these species significantly reduced litter decomposition, soil microbial biomass, and soil respiration on the large islands to levels more characteristic of small islands. In contrast, species removals on the small islands had no detectable effects on these properties, probably because of the increasing relative importance of abiotic drivers. This points to species' effects (and consequences of species losses) in ecosystems being context-dependent and of diminishing importance as retrogressive succession proceeds, as well as to the role of understory species in governing the effects of successional status on the functioning of the belowground subsystem. Any consequences of restoration effort for both the aboveground and belowground subsystems will therefore depend in a large part upon how the relative abundances of these understory species are influenced by frequency of fire regime.

The island system provides evidence that reducing fire frequency, and the ecosystem retrogression that results, greatly influences ecosystem C sequestration (Wardle *et al.* 2003). As retrogressive succession proceeds (i.e., as islands become smaller), standing plant biomass and NPP are both impaired, in both the tree and understory shrub layers. This results in less C storage aboveground. Further, decomposition rates of plant litter are reduced, for three reasons: (1) because plant species that produce poorer litter quality (*Picea* and *Empetrum*) begin to dominate; (2) because of phenotypic plasticity within plant species (i.e., plants of a given species producing poorer quality litter during retrogression); and (3) because of lower activity of the decomposer microflora. Further, during retrogression, decomposition is impaired before NPP, which results in a net gain of C to the soil. As such, the largest islands store less than 5 kg C/m^2 in the soil while some of the smallest ones store over 35 kg C/m^2. Because the belowground (rather than the aboveground) component stores the majority of C in these forests, there is net ecosystem C sequestration over time, in the order of 0.45 kg C/m^2 for every century without a major fire. These results point to fire suppression in forest ecosystems contributing significantly to C sequestration, and if these patterns are widespread elsewhere then this may play an important part in the global carbon cycle. They also point to the fact that ecosystem restoration involving the reintroduction or encouragement of natural fire regimes (and consequent reversion of retrogressive succession), would in turn alter the balance between aboveground and belowground processes, and thereby greatly influence total ecosystem carbon storage.

While there has been substantial fire suppression in the boreal forests of northern Scandinavia over the past couple of centuries, there is recent recognition that the use of fire in these systems may have beneficial consequences from a conservation perspective, and this issue has generated significant recent debate (Niklasson and Granström 2004). However, in Scandinavian forests, prescribed burning has been increasingly introduced in forests that are managed for production, mainly because of their perceived benefits for nutrient cycling and long-term forest stand productivity (Niklasson and Granström 2004, Nilsson and Wardle 2005). The studies described above for forested islands, as well as other related studies (reviewed by Nilsson and Wardle 2005) provide evidence that restoration of natural fire regimes (or implementation of prescribed burning) should serve to reverse ecosystem retrogression, promote rates of soil processes, and enhance forest productivity, at least in the order of decades. At a more globally relevant level, these forests are also important global carbon sinks or sources and restoration through introduction of a fire regime is likely to exert major effects on the amounts of carbon these forests store or release as CO_2. Most of the forests in Scandinavia are under some form of management and are utilized to varying degrees for timber and pulp production. Management for conservation benefits and for production forestry are not necessarily incompatible, and restoration of appropriate fire regimes may be useful for maximizing goals associated with each of these activities.

3.5 Belowground Impacts of Invasive Nonnative Plants

Invasive nonnative plants are widely perceived to have major, negative impacts in ecosystems (Pimentel et al. 2000, Mack et al. 2000, Myers and Bazley 2003), and the nature of spread and increased abundance of invasive plants has been documented for several species and systems. The relationships of these invaders with aboveground properties such as NPP and native vegetation diversity have been frequently studied (Daehler 2003, Ehrenfeld 2003, Levine et al. 2003), and it is recognized that these relationships vary strongly among both systems and species (see Hierro et al. 2005, Yurkonis et al. 2005; see Chapter 6). In contrast, impacts on belowground properties, processes, and communities are less well understood, although these may have profound implications for both successional processes and restoration efforts (Suding et al. 2004, Wolfe and Klironomos 2005, Young et al. 2005).

The best-documented belowground consequences of invasive plants are increases in soil nutrient fluxes, particularly of N. This is because many of the most widespread and successful invaders are plants associated with N-fixing symbionts, for example, those in the genera Acacia, Cytisus, Lupinus, Melilotus, Ulex, and Trifolium (Ehrenfeld 2003, Levine et al. 2003). These N-fixing plants should have important belowground impacts in N-limited ecosystems, as was first demonstrated by Vitousek et al. (1987) who found that the woody N-fixing invader Myrica faya increased N availability in Hawaiian ecosystems above that in native-dominated systems. Not surprisingly, many subsequent studies on the impacts of N-fixing invaders have found elevated levels of N availability across a range of species and systems (reviewed by Ehrenfeld 2003, Levine et al. 2003). A common assumption across these studies is that invading N-fixing plants are better able to garner N than their native counterparts, presumably due to their higher growth rates or greater per capita impacts on nutrient inputs into a system (although these possibilities have been little explored to date). A recent study by Weir et al. (2004) showed that different N-fixing bacterial mutualists are found in the root nodules of exotic plants from those found in native plant species, but it is unknown to what extent these novel root mutualists drive the high N inputs by invasive nonnative plants. Many non-N-fixing invasive species have higher growth rates, foliar nutrient contents, or litter inputs than do the native vegetation of the habitat being invaded, and can therefore also increase N availability in invaded ecosystems (e.g., Herman and Firestone 2005, Lindsay and French 2005).

Although most attention has focused on how plant invaders elevate N availability in ecosystems, there is increasing evidence that invaders also influence P availability and therefore the stoichiometry of systems. For example, the widespread invading shrub *Buddleja* has much higher foliar P than do other plant species dominating primary succession in both Hawaii (Matson 1990) and New Zealand (Bellingham *et al*. 2005). Similarly, *Pinus contorta* invading temperate grasslands increases soil P availability and rates of P cycling (Chen *et al*. 2003). Although the impacts of contrasting plant species on soil nutrient status are well documented (reviewed in Binkley and Giardina 1998, Wardle 2002, Ehrenfeld *et al*. 2005), the long-term implications of altered nutrient fluxes or stoichiometry for successional processes, ecosystem properties, and restoration are largely unknown. An unresolved issue is whether enhanced nutrient availability or terrestrial eutrophication caused by invaders has long-term negative impacts on late successional species composition, diversity, and successional trajectories (i.e., through promoting early successional species that are more nutrient-demanding). Results from ecosystem models suggest that nutrient inputs early in succession can have large and persistent effects on long-term productivity (e.g., Rastetter *et al*. 2003, Walker and del Moral 2003), exert differential impacts on later successional species (Bellingham *et al*. 2001), and influence species coexistence or persistence (Miki and Kondoh 2002).

Managers and restoration ecologists have sought to mitigate nutrient inputs from invaders by applying soil impoverishment treatments to reduce soil fertility levels (e.g., Alpert and Maron 2000, Wilson 2002, Corbin and D'Antonio 2004). This is typically accomplished by adding a carbon source to soils that can increase the soil microbial biomass and induce short-term reductions in nutrient availability to plants (e.g., Morghan and Seastedt 1999, Blumenthal *et al*. 2003, Yelenick *et al*. 2004). For example, additions of sucrose at 160 g m^{-2} yr^{-1} to sagebrush-prairie vegetation in Colorado increased plant species richness by an average of 2.1 species per 0.25 m^2 plot after 8 years of treatment (McLendon and Redente 1992). In contrast, Peltzer and Wilson (unpublished data) found that addition of C as sawdust at a rate of 400 g C m^{-2} yr^{-1} over 5 years to a prairie grassland restoration project in western Canada did not significantly decrease soil N or increase plant diversity. Adding C to soil is an extremely intensive undertaking and is likely to result in only short-term nutrient reduction. Overall, the evidence that soil impoverishment using carbon additions as a method for shifting the balance of species composition from exotic- to native-dominated systems or increasing plant diversity is weak (Corbin and D'Antonio 2004), and hence has not proven to be realistic for restoration management.

The impacts of invaders on the soil microbial community and soil fauna are less well understood than their impacts on nutrient availability. However, interactions between invasive plants and soil biota have been receiving increased recent attention in the literature for at least three reasons: soil communities can control the success of invaders in their new habitats (e.g., Reinhart *et al.* 2003, Callaway *et al.* 2004a and b); important shifts occur in belowground communities associated with plant invaders (e.g., Kourtev *et al.* 2003, Belnap *et al.* 2005, Yourkonis *et al.* 2005); and soil organisms can be strongly involved with invader impacts on native flora or ecosystem properties (e.g., Wolfe and Klironomos 2005). For example, soil pathogenic fungi or nematodes are implicated in the failure of new plant species to invade novel habitats or to subsequently spread (e.g., Mitchell and Power 2003, Reinhart *et al.* 2003, van der Putten 2005), whereas soil mutualists such as N-fixing bacteria or mycorrhizal fungi may promote the success or spread of invaders (e.g., Richardson *et al.* 2000, Klironomos 2002, Wolfe and Klironomos 2005). Plant invaders may increase ecosystem NPP through higher growth rates or per capita inputs of C and nutrients to the belowground subsystem than do the resident native species, thus resulting in higher soil microbial biomass and cascading benefits to other soil trophic levels such as bacterial and fungal feeding nematodes (Wardle 2002, Knevel *et al.* 2004). Further, soil organisms are implicated as either mediating or controlling invader impacts on other plant species, nutrient availability, or diversity. For example, the forb *Centaurea maculosa* has been shown to function as a successful, high-impact invader in western US grassland systems because of several mechanisms involving belowground communities including: sequestering P from neighboring plants via mycorrhizal fungi (Zabinski *et al.* 2002); suppressing native plants via allelopathic root exudates (Bais *et al.* 2003); and escaping soil pathogens and other enemies from its home range in eastern Europe and Asia (Callaway *et al.* 2004a and b, Hierro *et al.* 2005). These studies illustrate the critical role that belowground communities can have in the successful establishment, spread, and subsequent impacts of invasive plants on both native species and ecosystem properties (summarized in Table 3.2).

Feedbacks between the aboveground and belowground components of ecosystems have received increasing attention over the past decade, in part because of the critical role that belowground communities play in linking plant invaders to changes in resource availability, community composition, and ecosystem properties (Wardle 2002, Callaway *et al.* 2004b, Wolfe and Klironomos 2005). Feedbacks between plant species and soil communities have been developed as models for species coexistence or succession (e.g., Bever 2003, Packer and Clay 2004), and a working hypothesis that remains largely untested is that invaders may create different feedbacks with the belowground subsystem than do the native plant species that they displace (e.g., Callaway *et al.* 2004b). Feedbacks may also set systems along different successional trajectories if soil communities differentially facilitate or suppress later successional species, alter soil fertility levels, or induce vegetation switches (Wilson and Agnew 1992) and crossing of thresholds of ecosystem states (Suding *et al.* 2004). Differences in plant–soil feedbacks between rare and common plant species or between native and nonnative invasive species are only beginning to be appreciated, (e.g., van der Putten *et al.* 1993, Klironomos 2002), although the available evidence supports the idea that soil communities are an important, if previously neglected, driver of invader impacts.

Both the impacts of invaders on belowground properties and processes, and the feedbacks between the above- and belowground components of ecosystems, are highly relevant for understanding aboveground successional changes and

Table 3.2 Summary of some roles that soil biota play in three phases of plant invasion: naturalization, spread, and impact. These examples illustrate that strong links between aboveground and belowground communities can occur throughout the invasion process.

Phase of invasion	Published roles of soil biota
(1) Naturalization	Mycorrhizal fungi promote the establishment of an invader (Stampe and Daehler 2003).
	Soil biota have negative impacts in a species home range but not its introduced range (Callaway *et al.* 2004a and b, Reinhart *et al.* 2003).
	Different N fixing bacterial species are found in the root nodules of invasive plants compared to native plants (Weir *et al.* 2004).
	Soil pathogenic fungi or nematodes may determine successful establishment of plants (Knevel *et al.* 2004, Reinhart *et al.* 2003, Mitchell and Power 2003, van der Putten 2005).
(2) Spread	Mycorrhizal fungi promote the ongoing establishment of an invader (Stampe and Daehler 2003, Wolfe and Klironomos 2005) or control a plant's abundance within a community (Klironomos 2002).
	Soil pathogenic fungi or nematodes may determine successful spread of naturalized plants (Reinhart *et al.* 2003, van der Putten 2005).
	Feedbacks between the soil biota and an invasive plant may either promote or slow weed spread (Packer and Clay 2004).
(3) Impact	Mycorrhizal fungi can drive the impacts of an invader on aboveground biota (Wolfe and Klironomos 2005). For example, the weedy herb *Centaurea maculosa* can garner P from neighboring plants via mycorrhizal hyphae (Zabinski *et al.* 2002).
	Invasive plants can produce belowground allelochemicals that suppress native plants (Bais *et al.* 2003).
	High growth rates or litter quality of the invader causes increases in the soil microbial biomass and shifts in the soil biota resulting in enhanced nutrient availability (e.g., Belnap *et al.* 2005, Herman and Firestone 2005, Lindsay and French 2005).
	Feedbacks between the soil biota and an invasive plant may result in system-level shifts or crossing of an ecosystem threshold (Wardle 2002, Suding *et al.* 2004).

restoration efforts. The most common restoration technique involves carbon addition treatments but it has variable success and no clear links to the role of soil communities in the success or impact of invaders. There is tremendous scope in restoration for linking how invaders alter soil communities or processes to the long-term implications of invasions for vegetation succession and ecosystem processes. An emerging theme from current literature is that plants can influence soil communities that in turn regulate plant community composition or ecosystem processes. Clearly the role of belowground communities in influencing the spread and impacts of invasive plant species has important implications for succession and restoration that are only just beginning to be recognized. This area will provide a fertile arena for research in the coming decades.

3.6 Conclusions and Future Challenges

An emerging theme in the ecological literature over the past decade is that aboveground–belowground linkages can drive community composition, successional processes, and ecosystem development. Each of these processes is essentially a time-dependent outcome of these linkages, i.e., community composition shifts most rapidly, succession more slowly, and ecosystem development slowest of all. This conceptual model is highly relevant for ecosystem restoration, and highlights the importance of considering the belowground subsystem in both short- and long-term processes. The examples we provide above for the interactions of herbivores, fire, and invasions in these processes each emphasizes the importance of a combined aboveground–belowground approach to understanding vegetation succession, the ecological role of disturbance, and the restoration of ecological interactions and processes.

Some aspects of these aboveground and belowground interactions are clearly predictable: aboveground communities shift through succession and ecosystem development, from relatively rapidly growing, nutrient demanding plant species to slower-growing, nutrient-conserving species. Parallel changes occur in the soil subsystem, with shifts from domination by the bacterial-based to the fungal-based energy channel, from arbuscular-mycorrhizal to ectomycorrhizal communities, and from r-selected to K-selected soil fauna. These corresponding shifts in aboveground and belowground communities influence the nature of feedbacks between them, and this in turn has important consequences for ecosystem processes such as nutrient and C fluxes, NPP, and ecosystem C sequestration. However, at local scales, the outcome of many of these interactions is context-dependent and hence not easily predictable; therefore understanding these linkages may or may not be helpful for restoration efforts depending on whether the linkages of interest are predictable over temporal and spatial scales relevant for restoration. The challenge for future work is to understand which aboveground–belowground linkages are important for restoration within a given system, and in particular, understand how belowground processes both determine and respond to restoration treatments. Below we outline some of the key challenges and future research areas for developing closer links between aboveground–belowground interactions, succession, and restoration.

Belowground communities clearly play a pivotal role in successional processes, plant community composition, and ecosystem processes (as outlined in the sections above), but their role in restoration remains poorly understand. A major issue that can be resolved in the short-term is determining whether restoration efforts that focus solely on the aboveground community have the desired effects on the belowground community. Restoration practitioners typically focus on the aboveground components because these are visible and amenable to manipulation. For the immediate future, soil communities will not be widely manipulated or used to assess restoration success for the practical reason that understanding shifts in belowground communities is relatively difficult and requires quite specialized skills to identify or quantify soil biota. Clearly, research is needed to determine how closely restoration efforts aboveground are mirrored belowground; this can be accomplished in the first instance by the inclusion of soil biologists in monitoring restoration projects. Most restoration treatments involve either the addition of native species or the removal of nonnative species; these adaptive management experiments can be extremely informative if they include carefully designed controls (i.e., intact native systems and un-manipulated, invaded reference systems), and then compare both aboveground and belowground properties among these treatments. This approach can be used to resolve key questions such as whether removing a weed species from a system reverses the effects of that weed belowground, whether there are persistent effects on the soil biota or nutrient availability in the system, and whether these effects are minor compared to differences between intact indigenous systems and un-manipulated systems. Similarly, do manipulations of animal herbivore densities, either through restoration of native species or the reduction of pest species, also restore soil communities? Models, long-term experiments, and observational studies (such as across well defined chronosequences) can be also used to determine what the long-term consequences of restoration treatments are for successional processes or ecosystem properties, although these approaches are probably most useful for systems that are highly predictable.

Altering the aboveground or belowground community by removing herbivores or invasive plants may not always produce the desired result for succession or ecosystem properties (i.e., their impacts are not immediately reversed by removals) because of belowground legacies that may persist for a long time. For example, plant invaders interact strongly with belowground communities both directly through promoting NPP (higher C inputs), and influencing litter quality and nutrient inputs to the soil; and indirectly through interactions with native plants or species that show different plant–soil feedbacks. Many of these effects will have long-term implications for successional trajectories and ecosystem properties that cannot be mitigated by the removal of weeds, the most commonly used tool in restoration efforts. In addition to persistent effects or legacies of treatments, there will also be important lags in response to different components of the soil subsystem to aboveground manipulations. For example, if nutrient inputs into a system (e.g., from a N-fixing weed or atmospheric deposition) are offset by reducing these inputs or through soil impoverishment treatments, does the belowground system return to its previous state quickly or does it retain accumulated nutrients and switch to an alternative steady state dominated by early successional, nutrient-demanding species? Some components of the soil subsystem, such as the microbial biomass, will respond rapidly to aboveground treatments (i.e., weeks to months) and will be more sensitive indicators of changes in belowground processes to restoration treatments than will more slowly responding soil meso- and macrofauna (i.e., months to years) or pools of soil organic matter and nutrients (i.e., months to decades). The key unresolved issue is which long-term ecosystem properties or community trajectories in restoration are predictable from short-term shifts in aboveground–belowground interactions.

In summary, there are several critical areas for future research linking aboveground and belowground processes in succession to understanding restoration. First, are the shifts or trajectories in aboveground communities manipulated in restoration also mirrored belowground? Current studies document shifts in both aboveground and belowground communities, and clearly point to links between these components, but recognizing the importance of these links to succession, ecosystem development or restoration requires the inclusion of soil biologists alongside aboveground expertise. Examples of research that should provide insights into restoration include: (i) determining whether successful restoration aboveground also restores belowground communities, (ii) understanding whether soil mutualists or pathogens can be used to either promote native species or suppress nonnative species respectively, (iii) understanding in what situations soil manipulations (e.g., C additions) help or hinder ecosystem restoration, and (iv) identifying to what extent shifts in belowground processes during primary and secondary succession from field studies can be used to determine restoration success. The implications of these topics for restoration are obvious: soils and their biota are an important consideration for restoration plans, both in terms of understanding human-induced impacts on ecosystem properties, and also for the successful establishment and persistence of native species.

This is an exciting time for scientists and managers alike to better link aboveground and belowground systems to understand their collective influence on successional processes, ecosystem development, and ecosystem restoration. The central theme emerging from our review is that soil communities are intimately linked to aboveground properties and processes and vice-versa. These linkages may or may not be a boon for restoration efforts, but an improved understanding of them may also open up new avenues of investigation for ecosystem restoration through the recognition of the critical role belowground communities play throughout succession and ecosystem development.

References

Aerts, R., and Berendse, F. 1988. The effect of increased nutrient availability on vegetation dynamics in wet heathland. *Vegetatio* 76:63–69.

Alpert, P., and Maron J. L. 2000. Carbon addition as a countermeasure against biological invasion by plants. *Biological Invasions* 2:33–40.

Atkinson, I. A. E., and Greenwood, R. M. 1989. Relationship between moas and plants. *New Zealand Journal of Ecology* 12(supplement):67–96.

Augustine, D. J., and McNaughton, S. J. 1998. Ungulate effects on the functional species composition of plant communities: Herbivore selectivity and plant tolerance. *Journal of Wildlife Management* 62:1165–1183.

Bais, H. P., Vepachedu, R., Gilroy, S., Callaway, R. M., and Vivanco, J. M. 2003. Allelopathy and exotic plants: From genes to invasion. *Science* 301:1377–1380.

Bardgett, R. D. 2005. *The Biology of Soil: A Community and Ecosystem Approach.* Oxford: Oxford University Press.

Bardgett, R. D., and Wardle, D. A. 2003. Herbivore mediated linkages between aboveground and belowground communities. *Ecology* 84:2258–2268.

Bardgett, R. D., Wardle, D. A., and Yeates, G. W. 1998. Linking above-ground and below-ground interactions: How plant responses to foliar herbivory influence soil organisms. *Soil Biology and Biochemistry* 30:1867–1878.

Bellingham, P. J., Walker, L. R., and Wardle, D. A. 2001. Differential facilitation by a nitrogen fixing shrub during primary succession influences relative performance of canopy tree species. *Journal of Ecology* 89:861–875.

Bellingham, P. J., Peltzer, D. A., and Walker, L. R. 2005. Contrasting effects of dominant native and exotic shrubs on floodplain succession. *Journal of Vegetation Science* 16:135–142.

Belnap, J., Phillips, S., Sherrod, S., and Moldenke, A. 2005. Soil biota can change after exotic plant invasion: Does this explain ecosystem processes? *Ecology* 86:3007–3017.

Bever, J. D. 2003. Soil community feedback and the coexistance of competitors: Conceptual frameworks and empirical tests. *New Phytologist* 157:465–473.

Binkley, D., and Giardina, C. 1998. Why trees affect soils in temperate and tropical forests: The warp and woof of tree/soil interactions. *Biogeochemistry* 42:89–106.

Blumenthal, D. M., Jordan, N. R., and Russell, M. P. 2003. Soil carbon addition controls weeds and facilitates prairie restoration. *Ecological Applications* 13:605–615.

Bonan, G. B., and Shugart, H. H. 1989. Environmental factors and ecological processes in boreal forests. *Annual Review of Ecology and Systematics* 20:1–28.

Bradshaw, A. D., and Chadwick, M. J. 1980. *The Restoration of Land: The Ecology and Reclamation of Derelict and Degraded Land.* Los Angeles: University of California Press.

Brown, V. K., and Gange, A. C. 1990. Insect herbivory below ground. *Advances in Ecological Research* 20:1–58.

Burney, D. A., and Flannery, T. F. 2005. Fifty millennia of catastrophic extinctions after human contact. *Trends in Ecology and Evolution* 20:395–401.

Callaway, R. M., Thelen, G. C., Barth, S., Ramsey, P. W., and Gannon, J. E. 2004a. Soil fungi alter interactions between the invader *Centaurea maculosa* and North American natives. *Ecology* 85:1062–1071.

Callaway, R. M., Thelen, G. C., Rodriquez, A., and Hoben, W. E. 2004b. Soil biota and exotic plant invasion. *Nature* 427:731–733.

Chapin, F. S., Walker, B. H., Hobbs, R. J., Hooper, D. U., Lawton, J. H., Sala, O. E., and Tilman, D. 1997. Biotic control over the functioning of ecosystems. *Science* 277:500–504.

Chen, C. R., Condron, L. M., Sinaj, S., Davis, M. R., Sherlock, R. R., and Frossard, E. 2003. Effects of plant species on phosphorus availability in a range of grassland soils. *Plant and Soil* 256:115–130.

Coleman, D. C., Reid, C. P. P., and Cole, C. V. 1983. Biological strategies of nutrient cycling in soil systems. *Advances in Ecological Research* 13:1–55.

Compton, J. E., and Boone, R. D. 2000. Long term impacts of agriculture on soil carbon and nitrogen in New England forests. *Ecology* 81:2314–2330.

Conway, M. J. 1949. Deer damage in a Nelson beech forest. *New Zealand Journal of Forestry* 6:66–67.

Corbin, J. D., and D'Antonio, C. M. 2004. Can carbon addition increase competitiveness of native grasses? A case study from California. *Restoration Ecology* 12:36–43.

Daehler, C. C. 2003. Performance comparisons of co-occurring native and alien invasive plants: Implications for conservation and restoration. *Annual Review of Ecology, Evolution and Systematics* 34:183–211.

De Deyn, G. B., Raaijmakers, C. E., Zoomer, H. R., Berg, M. P., De Ruiter, P. C., Verhoef, H. A., Bezemer, T. M., and van der Putten, W. H. 2003. Soil invertebrate fauna enhances grassland succession and diversity. *Nature* 422:711–713.

De Deyn, G. B., Raaijmakers, C. E., van Ruijven, J., Berendse, F., and van der Putten, W. H. 2004. Plant species identity and diversity on different trophic levels of nematodes in the soil food web. *Oikos* 106:576–586.

De Luca, T. H., Nilsson, M.-C., and Zackrisson, O. 2002a. Nitrogen mineralization and phenol accumulation along a fire chronosequence in northern Sweden. *Oecologia* 133:206–214.

De Luca, T. H., Zackrisson, O., Nilsson, M.-C., and Sellstedt, A. 2002b. Quantifying nitrogen-fixation in feather moss carpets of boreal forests. *Nature* 419:917–920.

Díaz, S., Grime J. P, Harris, J., and McPherson, E. 1993. Evidence of a feedback mechanism limiting plant response to elevated carbon dioxide. *Nature* 364:616–617.

Dighton, J., and Mason, P. A. 1985. Mycorrhizal dynamics during forest tree development. In: *Developmental Biology of Higher Fungi*. D. Moore, L. A. Casselton, D. A. Wood, and J. C. Frankland (eds.). Cambridge: Cambridge University Press, pp. 117–139.

Duncan, K. W., and Holdaway, R. N. 1989. Footprint pressures and locomotion of moas and ungulates and their effects on the New Zealand indigenous biota by trampling. *New Zealand Journal of Ecology* 12(supplement):97–101.

Dunger, W., Schulz, H.-J., Zimdars, B., and Hohberg, K. 2004. Changes in collembolan species composition in eastern German mine sites over fifty years of primary succession. *Pedobiologia* 48:503–517.

Ehrenfeld, J. G. 2003. Ecosystem effects and causes of exotic species invasions. *Ecosystems* 6:503–523.

Ehrenfeld, J. G., Ravit, B., and Elgersma, K. 2005. Feedback in the plant-soil system. *Annual Review of Environment and Resources* 30:75–115.

Eviner, V. T., and Chapin, F. S., III. 2003. Functional matrix: A conceptual framework for predicting multiple plant effects on ecosystem processes. *Annual Review of Ecology and Systematics* 34:455–485.

Forsyth, D. M., Coomes, D. A., Nugent, G., and Hall, G. M. 2002. Diet and diet preferences of introduced ungulates (order: Artiodactyla) in New Zealand. *New Zealand Journal of Zoology* 29:323–343.

Forsyth, D. M., Richardson, S. J., and Menchenton, K. 2005. Foliar fibre predicts diet selection by invasive Red Deer (*Cervus elaphus scoticus*) in a temperate New Zealand forest. *Functional Ecology* 3:495–504.

Grime, J. P. 1979. *Plant Strategies and Vegetation Processes*. Chichester: Wiley.

Hamilton, E. W., and Frank, D. A. 2001. Can plants stimulate soil microbes and their own nutrient supply? Evidence from a grazing tolerant grass. *Ecology* 82:2397–2402.

Herman, D. J., and Firestone, K. 2005. Plant invasion alters nitrogen cycling by modifying the soil nitrifying community. *Ecology Letters* 8:976–985.

Hierro, J. L., Maron, J. L., and Callaway, R. M. 2005. A biogeographical approach to plant invasions: The importance of studying exotics in their introduced and native range. *Journal of Ecology* 93:5–15.

Hodkinson, I. D., Coulson, S. J., and Webb, N. R. 2004. Invertebrate community assembly across proglacial chronosequences in the high Arctic. *Journal of Animal Ecology* 73:556–568.

Hurtt, G. C., Pacala , S. W., Moorcroft, P. R., Caspersen, J., Shevliakova, E., Houghton, R., and Moore, B. 2002. Projecting the future of the US carbon sink. *Proceedings of the National Academy of Sciences, U.S.A.* 99:1389–1394.

Klironomos, J. N. 2002. Feedback with soil biota contributes to plant rarity and invasiveness in communities. *Nature* 417:67–70.

Knevel, I. C., Lans, T., Menting, F. B. J., Hertling, U. M., and van der Putten, W. H. 2004. Release from native root herbivores and biotic resistance by soil pathogens in a new habitat both affect the alien *Ammophila arenaria* in South Africa. *Oecologia* 141:502–510.

Kourtev, P. S., Ehrenfeld, J. G., and Haggblom, M. 2003. Experimental analysis of the effect of exotic and native plant species on the structure and function of soil microbial communities. *Soil Biology and Biochemistry* 35:895–905.

Laakso, J., and Setälä, H. 1999. Sensitivity of primary production to changes in the architecture of belowground food webs. *Oikos* 87:57–64.

Lawton, J. H. 1994. What do species do in ecosystems? *Oikos* 71:367–374.

Levine, J. M., Vila, M., D'Antonio, C. M., Dukes, J. S., Grigulis, K., and Lavorel, S. 2003. Mechanisms underlying the impacts of exotic plant invasions. *Proceedings of the Royal Society of London - Series B* 270:775–781.

Liiri, M., Setälä, H., Haimi, J., Pennanen, T., and Fritze, H. 2002. Relationship between soil microarthropod species diversity and plant growth rates does not change when the system is disturbed. *Oikos* 96:138–150.

Lindsay, E. A., and French, K. 2005. Litterfall and nitrogen cycling following invasion by *Chrysanthemoides monilifera* ssp. *rotundata* in coastal Australia. *Journal of Applied Ecology* 42:556–566.

Luken, J. O., and Fonda, R. W. 1983. Nitrogen accumulation in a chronosequence of red alder communities along the Hoh River, Olympic National Park, Washington. *Canadian Journal of Forest Research* 13:1228–1237.

Mack, R. N., Simberloff, D., Lonsdale, W. M., Evans, J., Clout, M., and Bazzaz. F. A. 2000. Biotic invasions: Causes, epidemiology, global consequences, and control. *Ecological Applications* 10:689–710.

Matson, P. A. 1990. Plant-soil interactions in primary succession at Hawaii Volcanoes National Park. *Oecologia* 85:241–246.

McGlone, M. S., and Clarkson, B. D. 1993. Ghost stories: Moa, plant defenses and evolution in New Zealand. *Tuatara* 32:1–21.

McLendon, T., and Redente, E. F. 1992. Effects of nitrogen limitation on species replacement dynamics during early succession on a semiarid sagebrush site. *Oecologia* 91:312–317.

McNaughton, S. J. 1985. Ecology of a grazing system: The Serengeti. *Ecological Monographs* 55:259–294.

McNaughton, S. J., Oesterheld, M., Frank, D. A., and Williams, K. J. 1989. Ecosystem-level patterns of primary productivity and herbivory in terrestrial habitats. *Nature* 341:142–144.

Miki, T., and Kondoh, N. 2002. Feedbacks between nutrient cycling and vegetation predict plant species coexistence and invasion. *Ecology Letters* 5:624–633.

Mitchell, C. E., and Power, A. G. 2003. Release of invasive plants from fungal and viral pathogens. *Nature* 421:625–626.

Morghan, K. J. R., and Seastedt, T. R. 1999. Effects of soil nitrogen reduction on non-native plants in restored grasslands. *Restoration Ecology* 71:51–55.

Morris, W. F., and Wood, D. M. 1989. The role of lupine in Mt. St. Helens: Facilitation or inhibition? *Ecology* 70:697–703.

Myers, J. H., and Bazely, D. R. 2003. *Ecology and Control of Introduced Plants*. Cambridge: Cambridge University Press.

Niklasson, M., and Granström, A. 2000. Numbers and sizes of fires: Long term spatially explicit fire history in a Swedish boreal landscape. *Ecology* 81:1484–1499.

Niklasson, M., and Granström, A. 2004. Fire in Sweden – history, research, prescribed burning and forest certification. *International Forest Fire News* 30:80–83.

Nilsson, M. C., and Wardle, D. A. 2005. Understory vegetation as a forest ecosystem driver: Evidence from the northern Swedish boreal forest. *Frontiers in Ecology and the Environment* 3:421–428.

Odum, E. P. 1969. The strategy of ecosystem development. *Science* 164:262–270.

Ohtonen, R., Fritze, H., Pennanen, T., Jumpponen, A., and Trappe, J. 1999. Ecosystem properties and microbial community changes in primary succession on a glacier forefront. *Oecologia* 119:239–246.

Packer, A., and Clay, K. 2004. Development of negative feedback during successive growth cycles of black cherry. *Proceedings of the Royal Society of London* 271:317–324.

Pastor, J., Dewey, B., Naiman, R. J., McInnes, P. F., and Cohen, Y. 1993. Moose browsing and soil fertility in the boreal forests of Isle Royale National Park. *Ecology* 74:467–480.

Pimentel, D., Lach, L., Zuniga, R., and Morrison, D. 2000. Environmental and economic costs of nonindigenous species in the United States. *Bioscience* 50:53–65.

Porazinska, D. L., Bardgett, R. D., Blaauw, M. B., Hunt, H. W., Parsons, A. N., Seastedt, T. R., and Wall, D. H. 2003. Relationships at the aboveground-belowground interface: Plants, soil biota and soil processes. *Ecological Monographs* 73:377–395.

Rastetter, E. B., Aber, J. D., Peter, D. P. C., Ojima, D. S., and Burke, I. C. 2003. Using mechanistic models to scale ecological processes across space and time. *BioScience* 53:68–76.

Reinhart, K. O., Packer, A., van der Putten, W. H., and Clay, K. 2003. Escape from natural soil pathogens enables a North American tree to invade Europe. *Ecology Letters* 6:1046–1050.

Richardson, D. M., Allsopp, N., D'Antonio, C. M., Milton, S. J., and Rejmánek, M. 2000. Plant invasions – the role of mutualisms. *Biological Reviews* 75:65–93.

Sankaran, M., and Augustine, D. 2004. Large herbivores suppress decomposer abundance in a semi-arid grazing ecosystem. *Ecology* 85:1052–1061.

Schimel, D. S. 1995. Terrestrial ecosystems and the carbon cycle. *Global Change Biology* 1:77–91.

Schlesinger, W. H., Bruijnzeel, L. A., Bush, M. B., Klein, E. M., Mace, K.A., Raikes J A., and Whittaker, R. J. 1998. The biogeochemistry of phosphorus after the first century of soil development on Rakata Island, Krakatau, Indonesia. *Biogeochemistry* 40:37–55.

Seastedt, T. R. 2000. Soil fauna and controls of carbon dynamics: Comparisons of rangelands and forests across latitudinal gradients. In: *Invertebrates as Webmasters in Ecosystems*. P. F. Hendrix and D. C. Coleman (eds.). Wallingford: CAB International, pp. 293–312.

Sigler, W. V., and Zeyer, J. 2002. Microbial diversity and activity along the forefields of two receding glaciers. *Microbial Ecology* 43:397–407.

Stampe, E. D., and Daehler, C. C. 2003. Mycorrhizal species identity affects plant community structure and invasion: A microcosm study. *Oikos* 100:362–372.

Suding, K. N., Gross, K. L., and Houseman, G. R. 2004. Alternative states and positive feedbacks in restoration ecology. *Trends in Ecology and Evolution* 19:46–53.

Van der Heijden, M., Klironomos, J. N., Ursic, M., Moutoglis, P., Streitwolf-Engel, R., Boller, T., Wiemkin, A., and Sanders, I. R. 1998. Mycorrhizal fungal diversity determines plant biodiversity, ecosystem variability and productivity. *Nature* 396:69–72.

van der Putten, W. H. 2005. Plant-soil feedback and soil biodiversity affect the composition of soil communities. In: *Biological Diversity and Function in Soils*. R. D. Bardgett, M. B. Usher, and D. W. Hopkins (eds.). Cambridge: Cambridge University Press, pp. 250–272.

van der Putten, W. H., Van Dijk, C., and Peters, B. A. M. 1993. Plant specific soil borne diseases contribute to succession in foredune vegetation. *Nature* 363:53–56.

van der Putten, W. H., Vet, L. E. H., Harvey, J. H., and Wackers, F. L. 2001. Linking above- and belowground multitrophic interactions of plants, herbivores, pathogens and their antagonists. *Trends in Ecology and Evolution* 16:547–554.

Verhoeven, R. 2002. The structure of the microtrophic system in a development series of dune sands. *Pedobiologia* 46:75–89.

Vitecroft, M., Palmborg, C., Sohlenius, B., Huss-Danell, K., and Bengtsson, J. 2005. Plant species effects on soil nematode communities in experimental grasslands. *Applied Soil Ecology* 30:90–103.

Vitousek, P. M. 2004. *Nutrient Cycling and Limitation: Hawai'i as a Model System*. Princeton: Princeton University Press.

Vitousek, P. M., Walker, L. R., Whiteaker, L. D., Mueller-Dombois, D., and Matson, P. A. 1987. Biological invasion by *Myrica faya* alters ecosystem development in Hawaii. *Science* 238:802–804.

Walker, L. R., and Chapin, F. S., III. 1987. Interactions among processes controlling successional change. *Oikos* 50:131–135.

Walker, L. R., and del Moral, R. 2003. *Primary Succession and Ecosystem Rehabilitation*. Cambridge: Cambridge University Press.

Walker, T. W., and Syers, J. K. 1976. The fate of phosphorus during pedogenesis. *Geoderma* 15:1–19.

Wallis, F. P., and James, I. L. 1972. Introduced animal effects and erosion phenomena in the northern Urewera forest. *New Zealand Journal of Forestry* 17:21–36.

Wardle, D. A. 2002. *Communities and Ecosystems: Linking the Aboveground and Belowground Components*. Princeton: Princeton University Press.

Wardle, D. A., and Bardgett, R. D. 2004. Human-induced changes in densities of large herbivorous mammals: Consequences for the decomposer subsystem. *Frontiers in Ecology and the Environment* 2:145–153.

Wardle, D. A., and Zackrisson, O. 2005. Effects of species and functional group loss on island ecosystem properties. *Nature* 435:806–810.

Wardle, D. A., Zackrisson, O., Hörnberg, G., and Gallet, C. 1997. Influence of island area on ecosystem properties. *Science* 277:1296–1299.

Wardle. D. A., Barker, G. M., Bonner, K. I., and Nicholson, K. S. 1998. Can comparative approaches based on plant ecophysiological traits predict the nature of biotic interactions and individual plant species effects in ecosystems? *Journal of Ecology* 86:405–420.

Wardle, D. A., Bonner, K. I., Barker, G. M., Yeates, G. W., Nicholson, K. S., Bardgett, R. D., Watson, R. N., and Ghani, A. 1999. Plant removals in perennial grassland: Vegetation dynamics, decomposers, soil biodiversity and ecosystem properties. *Ecological Monographs* 69:535–568.

Wardle, D. A., Barker, G. M., Yeates, G. W., Bonner, K. I., and Ghani, A. 2001. Introduced browsing mammals in natural New Zealand forests: Aboveground and belowground consequences. *Ecological Monographs* 71:587 614.

Wardle, D. A., Bonner, K. I., and Barker, G. M. 2002. Linkages between plant litter decomposition, litter quality, and vegetation responses to herbivores. *Functional Ecology* 16:585–595.

Wardle, D. A., Hörnberg, G., Zackrisson, O., Kalela-Brundin, M., and Coomes, D. A. 2003. Long term effects of wildfire on ecosystem properties across an island area gradient. *Science* 300:972–975.

Wardle, D. A., Bardgett, R. D., Klironomos, J. N., Setälä, H., van der Putten, W. H., and Wall, D. H. 2004a. Ecological linkages between aboveground and belowground biota. *Science* 304:1629–1633.

Wardle, D. A., Walker, L. R., and Bardgett, R. D. 2004b. Ecosystem properties and forest decline in contrasting long-term chronosequences. *Science* 305:509–513.

Wasilewska, L. 1994. The effect of age of meadows on succession and diversity in soil nematode communities. *Pedobiologia* 38:1–11.

Weir, B. S., Turner, S. J., Silvester, W. B., Park, D.-C., and Young, J. M. 2004. Unexpectedly diverse *Mesorhizobium* strains and *Rhizobium leguminosarum* nodulate native legume genera of New Zealand, while introduced legume weeds are nodulated by *Bradyrhizobium* species. *Applied and Environmental Microbiology* 70:5980–5987.

Wilson, J. B., and Agnew, A. D. Q. 1992. Positive-feedback switches in plant communities. *Advances in Ecological Research* 23:263–336.

Wilson, S.D. 2002. *Prairies. Handbook of Ecological Restoration.* A. J. Davy and M. R. Perrow (eds.). Cambridge University Press, Cambridge, pp. 443–465.

Wolfe, B. E., and Klironomos, J. N. 2005. Breaking new ground: Soil communities and exotic plant invasion. *BioScience* 55:477–487.

Yelenik, S. G., Stock, W. D., and Richardson, D. M. 2004. Ecosystem level impacts of invasive *Acacia saligna* in the South African fynbos. *Restoration Ecology* 12:44–51.

Yurkonis, K. A., Meiners, S. J., and Wachholder, B. E. 2005. Invasion impacts diversity through altered community dynamics. *Journal of Ecology* 93:1053–1061.

Young, T. P., Petersen, D. A., and Clary, J. J. 2005. The ecology of restoration: Historical links, emerging issues and unexplored realms. *Ecology Letters* 8:662–673.

Zabinski, C.A., Quinn, L., and Callaway, R.M. 2002. Phosphorus uptake, not carbon transfer, explains arbuscular mycorrhizal enhancement of *Centaurea maculosa* in the presence of native grassland species. *Functional Ecology* 16:758–765.

Zackrisson, O., Nilsson, M.-C., and Wardle, D. A. 1996. Key ecological function of charcoal from wildfire in the boreal forest. *Oikos* 77:10–19.

Zimov, S. A., Chuprynin, V. I., and Oreshko, A. P. 1995. Steppe–tundra transition: A herbivore-driven biome shift at the end of the Pleistocene. *The American Naturalist* 146:765–794.

4

Retrogressive Succession and Restoration on Old Landscapes

Joe Walker and Paul Reddell

Key Points

1. Natural retrogressive succession is common on old, highly weathered landscapes in many parts of the world.
2. Disturbance in such areas results in an acceleration of the regressive state and this limits restoration potential and end-point definition.
3. Ameliorating of soil physical properties, reinvigorating soil biology, modifying the water cycle, and applying suitable quantities of missing nutrients to a site are key considerations in developing restoration strategies for such landscapes.
4. In many areas there may be landscape "analogues" of the target restored state, where natural erosion and weathering have progressed more rapidly in particular landscape locations and where existing plant communities can be used as models for species composition, structure, productivity, and function in the restoration process.

4.1 Introduction

Any environmental restoration program should progress through stages of planning, implementation, and monitoring. It is mainly in the planning phase that concepts about succession are most useful. In this phase, a key conceptual question relates to whether ecosystems developed on old weathered landscapes have the same resilience (the ability to spring back) as young landscapes developed after the last glacial period or on new substrates. Many old landscapes are known to exhibit natural retrogression; they become leaky, losing nutrients or become less efficient in carbon or water cycling. Consequently, old landscapes can be expected to respond differently to disturbance and restoration activities compared with young landscapes (Walker *et al.* 2001). Hence the identification of the end-points of a restoration program could differ markedly between young and old landscapes. Knowledge about the functional differences between young and old landscapes can provide information needed to facilitate a paradigm shift in the approach to the restoration of ecosystems in old landscapes. We will use examples from semiarid and tropical systems to demonstrate the need for detailed process knowledge about succession when trying to restore old landscapes following land-use disturbances.

4.2 Natural Retrogressive Succession

Succession theory in ecology had its beginnings in studies of vegetation changes across sand dune chronosequences (Warming 1895, Cowles 1895). A progressive build up to a maximum biomass and species richness were observed across the dunes, and these changes were accompanied by gradual changes in the physicochemical composition of the soils. The early vegetation work was contemporary with the idea of geographical cycles proposed by Davis (1899) to explain how landforms evolved over time, and his schema included renewal and degradation. The vegetation observations led to the idea of a "climax community," that is, the progressive development toward an optimum expression of vegetation for the climate and soils of a region. As ecological ideas developed, the climax state was seen to fluctuate but it was generally regarded as in a stable or dynamic equilibrium state. In terms of restoration actions, succession trajectories toward the original "climax" state have been viewed as desirable. Debate continued about the dynamics and trajectory of successional processes, but gradually it was realized that natural post-climax states exist. Early studies using pollen analysis in peat or mor layers in postglacial deposits (Iversen 1964) showed examples of ecosystems with permanently reduced productivity. These so-called retrogressive successions were associated with leaching of the soils during pedogenesis (natural retrogression) and man-made disturbances (secondary retrogression). A clear demonstration of a post-climax natural retrogression was described for an intact aeolian sand dune chronosequences at Cooloola, Queensland, Australia by Walker *et al.* (1981) and Thompson (1981 and 1983). Dune building at Cooloola was episodic, and some parts of earlier dune systems were not buried by subsequent wind-blown deposits. These exposed parts were subjected to weathering, leaching, and erosion, and form an age sequence that stretches over some 750K years (Thompson 1992). These studies showed that for freely drained sites vegetation type, species richness, standing biomass, and soil carbon accumulation varied with dune age. After a period of biomass build up (progressive succession) seen in the youngest four dune systems, vegetation biomass and the store of organic material in the surface soils declined in the oldest three systems (retrogressive succession). Vitousek and Reiners (1975), Hedin *et al.* (2003) and Wardle *et al.* (2004) have shown for a series of very long-term chronosequences, that ecosystem decline is a widespread phenomenon. However, the mechanisms controlling retrogression vary between bioregions and with disturbance regimes and intensity.

The hypothesis developed from the Cooloola study (Fig. 4.1) relevant to planning restoration actions is that pedologically young landscapes when disturbed tend to recover toward the previous state, whereas old systems become "leaky" or cannot recover critical system functions and trend toward a new system state with lower biomass and complexity.

Old landscapes are common in Australia but also occur in most parts of the southern hemisphere and in tropical areas. At a global scale old landscapes predominate, yet many of the developments in succession theory were carried out in pedologically young landscapes, and perhaps the same is true of attempts to develop links with restoration activities. Like many areas in the tropics, Australia

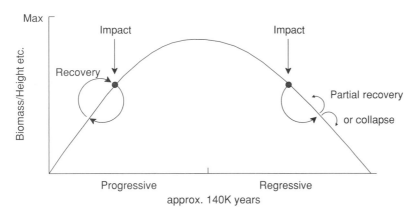

Figure 4.1 In young landscapes after an impact the vegetation tends to return to the original state, whereas on old landscapes vegetation will only partially recover or collapse (modified from Walker *et al.* 2001).

escaped the direct effects of the Pleistocene glaciations that destroyed the Tertiary landscapes and weathering mantles over much of the northern hemisphere. In old landscapes, the residual products of weathering have formed thick soil mantles generally lacking weatherable minerals and have low concentrations of available plant nutrients. The extensive areas of lateritic soils in Australia, and in the tropics generally are examples of this (Sanchez 1976). As secondary minerals form in the weathering zone, there is an initial increase in available plant nutrients, but this decreases rapidly as the supply of weatherable minerals is depleted (Stark 1978, Walker and Syers 1976). With progressive weathering, plant nutrient availability decreases, and plant species trend toward those with more efficient mechanisms of nutrient capture and storage. Aboveground biomass is reduced, the proportion of belowground biomass increases and physiological and morphological adaptations are common (Lamont 1981, Walker *et al.* 1987, Pate 1994). The system may be stable in its undisturbed state, but is particularly vulnerable to disturbance as shown by a lesser ability to recycle nutrients or retain water. An example is the failure of tropical rainforests on yellow podzolic or lateritic soils to regenerate after clearing, burning, and agricultural use (Sanchez 1976). The implication is that halting an inevitable decline after disturbance may be impossible in old landscapes, and at best one has to settle for a stable system state different from the original as a restoration end-point.

In summary, natural retrogressive succession has the following attributes:

1. A decline in ecosystem productivity and complexity occurs over millennia due to processes associated with soil weathering, especially a decline in nutrient availability and declines in a range of soil properties that affect the water cycle.
2. The intact ecosystems of old landscapes are usually stable as the result of the gradual development of adaptive plant physiological and morphological traits.
3. Man-induced disturbances (clearing, burning, grazing, and farming) push old systems quickly into further decline (new system states) and key system functions may be lost permanently.
4. Restoration expectations (end-points) in old landscapes are different from young landscapes and a stable system state different from the original state may be the best outcome.

5. Restoring the hydrological properties of degraded old landscapes to reduce water and nutrient leakage is a critical first step. This places the focus on acquiring and using knowledge about the functional characteristics of plants, for example, plant rooting strategies to improve resource availability, and understanding mechanisms about changes to the water cycle and the role of soil biota.

The following are examples of restoration issues on old landscapes from a semiarid ecosystem and a tropical ecosystem where recognition and understanding of retrogressive succession may have a profound impact on the success and effectiveness of restoration activities.

4.3 Human-Induced Salinization in Southern Australia: A Symptom of Changes to the Water Cycle of Old, Semiarid Landscapes

Across southern Australia, landscape degradation issues are dominated by the widespread occurrence of secondary salinity (salinity that appears after disturbance), water logging, soil acidification, and soil structural decline. In combination, they greatly affect water quality in streams, terrestrial and aquatic biodiversity, roads and other infrastructure, and reduce ecosystem goods and services (Hatton 2002). Areas with a Mediterranean climate, characterized by wet winters with low evaporation rates, are particularly affected (McFarlane and Williamson 2002). All the degradation issues have a common link—changes to the inputs, outputs, and storage of water following the removal of the native vegetation for human activity. Secondary salinization is considered here as one aspect of the inability of old landscapes to sustain productivity and recover following broad-scale vegetation changes and in many cases inappropriate land uses.

Salt-affected soils occur extensively across Australia, and are estimated to cover approximately one-third of the continent (Northcote and Skene 1972, NLWRA 2001). The origins of salt in weathered landscapes, halomorphic soils, and salt lakes in Australia have been attributed first to long-term weathering of rocks of marine or lacustrine origin (Blackburn 1976), and second to atmospheric accessions of inorganic aerosols in rain or dry fallout of either ocean or terrestrial origin over geological time (Allison *et al.* 1983, Herczeg *et al.* 2001). The presence of appreciable amounts of salt in deeply weathered landscapes is considered to be a relict feature because saline soils are surrounded by nonsaline soils developed from fresh rocks (Gunn and Richardson 1979). Salinization of landscapes thus occurred prior to denudation and has been supplemented by atmospheric and aeolian inputs since these earlier geological times. Even before European settlement in Australia in the late 18th century, salty outbreaks in woodlands and forested landscapes were part of Australia's environment.

In arid and semiarid Australia, evaporation considerably exceeds rainfall. These areas are also characterized by low slopes with low hydraulic gradients and hence a lack of lateral water movement. Vertical leaching and little lateral movement have resulted in subsoil salt stores. Much of the salt is stored below the root zone of trees. To become a problem for agricultural production or to pollute streams, the stored salt has to be mobilized vertically or laterally. Since European settlement in the late 18th century, extensive agricultural land-use following tree clearing has induced major changes to soil hydraulic properties of the thin soil mantle, and hence the water cycle generally. With less evapotranspiration resulting from vegetation clearing, more water percolates beyond the

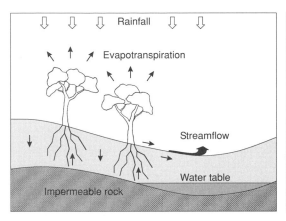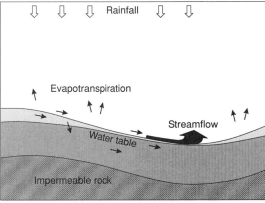

Figure 4.2 Generalized water inputs, outputs and flows in a treed versus cleared landscape showing water table rises, salt mobilization, and increased stream flow. The scale can vary from a hill slope (100s of meters) to a region (100s of kilometers).

root zone into water tables and accumulates. With increasing stored groundwater, water tables subsequently rise and mobilize salt upward into the root zone of plants (Fig. 4.2). Rising water tables can result from local or regional scale groundwater accumulations.

If left unchecked, retrogressive succession moves the system toward vegetation collapse and development of bare salinized soil areas (Fig. 4.3). In addition, secondary toxic mineral compounds can be produced and enter stream networks (Fitzpatrick *et al.* 1996). The consequences are that land salinization and water logging cost rural producers in the order of hundreds of millions of dollars in

Figure 4.3 Tree clearing on the upper slopes has caused salinization of the lower slopes and the drainage line resulting in loss of ground cover and soil erosion (photo by John Gallant, CSIRO).

production per year as well as impacting on roads, buildings, and other infrastructures. About 10% of the rural lands in Western Australia are affected by salinization and this figure could double in the next few decades (George *et al.* 1997). In southeastern Australia about 5% of rural lands are salt affected (NL-WRA 2001). Generally, salinization resulting from local groundwater systems has a lesser impact than those from regional systems.

The broad impacts of land-use on the water cycle across landscapes since European settlement are summarized by Williams *et al.* (2001) as follows:

$$\text{Rainfall (P)} = \text{ET} + \text{DD} + \text{RO} + \text{IF} + \text{GWD} + \text{SBWS}$$

- ET – evapotranspiration has decreased due to lower leaf area (LAI) values of the vegetation when summed on a yearly basis (trees replaced by annual crops)
- DD – deep drainage (recharge or leaching) has increased as shown by increasing groundwater pressures in salinized areas, but the change in water amount is a small fraction of the total water balance and is hard to measure or model
- RO – surface run-off (streamflow) has increased due to soil structural changes that reduce soil permeability and surface soil storing capacity
- IF – interflow water that moves laterally down slope within the surface soil has decreased due to a shallower soil A-horizon and soil structural decline
- GWD – ground water discharge has increased as evidenced by increased areas of salinity and waterlogging
- SBWS – soil water storage has decreased in the unsaturated soil profile (reduced in depth due to erosion and less soil organic matter) and less biological water is stored in vegetation—that is, the surface soils are drier overall.

The key mechanisms operating in these old systems that originally stored water and maintained soil profile integrity have been disrupted. The overall changes in the water cycle following tree removal are a shift toward desertification, i.e., a drier landscape, which is a consequence that has yet to be widely accepted. That disturbed landscapes are drier than the original presents an apparent paradox—drier surface soils but rising water tables. Williams *et al.* (2001) suggest several mechanisms to explain this apparent paradox. These include an increase in preferred water flow pathways at a range of scales from micropores to hill slopes (termed holeyness) and a reduced capacity to store water due to shallower soil A-horizons. Extensive tree planting to reduce the area affected by salinity is widely adopted in Australia as a restoration method but at least initially, this tends to dry out the landscapes even more. An alternative is to develop restoration strategies that increase storage of soil moisture at the surface and release water slowly to the root zone.

Given the context is restoration within salinized agriculturally productive landscapes, the desirable end points of restoration actions can be stated as:

1. A mosaic of vegetation types that restore the hydrological functioning of the soil and at the same time maintain economic viability.
2. Improved goods and services from the production landscape—water quality and quantity with less export of saline water, and increased wildlife habitat.

What knowledge is needed and what tools are available to restore salinized landscapes to more productive agricultural land? We need to know how to manipulate rainfall absorption at a range of scales and understand the magnitude and location of water and energy flows through the landscape, including the coupling with groundwater. Research approaches include identification and quantifying the main system drivers and various forms of ecohydrological modeling. Modeling how water moves in and through a heterogeneous landscape and how it is modified by various management or restoration actions is extremely difficult due to inherent spatial variability in soil properties (Hatton 2002, Williams *et al.* 2006). Soil tracer methods are useful in small uniform plots but suffer from spatial variability. A number of process-based ecohydrological models are available to simulate how changed land-use and restoration affects ground water recharge, for example, the WAVES model Zhang and Dawes (1998) and Silberstein *et al.* (1999). Other relevant models that range in spatial scales from plots to regions and with various levels of complexity are reviewed by Walker *et al.* (2002). But one has to accept that these process-based approaches involve broad approximations, and because of the lack of lateral flow across the polygons or cells of most models, extrapolation beyond the research plot is difficult. The most appropriate use of process-based modeling involves the development of possible scenarios that are amenable to field verification. Likewise, the partitioning of rainfall excess into various forms of predicted outflow is highly uncertain given variable patterns of plant water use and seasonal differences in rainfall (Dunin 2002). Spatial heterogeneity exists at many scales from preferred pathways of water through soils at a given site to patterns at regional scales (Williams *et al.* 2001). At the broadest scales, the overall control of salinization is through topography, because in a general sense topography integrates lateral and vertical water movement. This link with topography is used in the FLAG model (Summerell *et al.* 2000) using detailed terrain data and a fuzzy logic approach to identify areas of potential salinization.

Given that the key mechanisms changed by agricultural development are associated with the surface soil layers, we suggest that the principal restoration objectives in salinized lands at the farm scale should be: First, develop a buffer or series of buffers in the surface soil that store and slowly release water. Practices that improve soil organic matter content and improve soil fauna activity would appear to be instrumental in developing such a buffer. Second, ensure that excess water is of high quality and is able to move to where it can be utilized or stored. Third, develop the agricultural landscape around a vegetation mosaic based on different functionality, including phases of succession from grassland to shrubland to woodland. In this scenario the vegetation mix includes pasture, crop, shrub, and tree species with various rooting depths and architecture selected to utilize water and nutrient resources from various soil layers.

The question now posed is: do any currently used restoration approaches meet these broad objectives?

4.4 Restoration Methods Used in Australia in Salt-Affected Landscapes

4.4.1 Widespread Tree Planting for Salinity Management

At the broad scale, the thinking that underpins the vast majority of investments in research and development programs, community projects, or commercial forestry, focuses on a tree planting solution. The idea is that deforestation caused widespread salinization, hence reforestation will solve the problem (Hatton and Nulsen 1999). The goal in tree planting is to reduce leakage to the ground water system. A major limitation is that secondary salinity is mainly present in areas of less than 550 mm annual rainfall and areas where winter rains dominate. The rate of evapotranspiration to dry out the soil is a function of leaf area index (LAI) and rooting depth. LAI levels sufficient to achieve this outcome are generally confined to areas with rainfalls greater than 600 mm and the best tree growth is in areas of greater than 1000 mm. Many reforestation strategies have little impact on rates of secondary salinization or salt loads into streams, and field trials suggest that some 30–50% of a catchment needs to be reforested to be effective (Hatton and Nulsen 1999, Salama and Hatton 1999). A high percentage of trees in the landscape are not useful in crop producing areas. Experience shows that the consequence of planting large areas to trees in semiarid lands is significant tree death (Walker and Nicholl 1996) or decreased stream flow (Zhang *et al.* 2001). Planting trees in discharge areas (areas where water logging or salty discharges are evident) has been largely ineffective because trees initially use up available water and soon succumb to high salt concentrations due to a lack of sufficient leaching in the root zone (Thorburn *et al.* 1995, Thorburn 1997). The exception to this is at small scales where there is only local ground water or perched water tables and where salt concentrations in the ground water are low: dense plantations of trees or shrubs (high LAI) can reduce the water table in these cases (Bell *et al.*1990, Greenwood *et al.*1992, Walker *et al.* 2002). Even at small scales, the extent of drying out is quite localized (in the range of 20–50 m from the tree edge; Ellis *et al.* 2005), tree survival can be poor, and impacts on stream salinity minimal (George *et al.* 1999, Walker and Nicholl 1996). Salt-tolerant plants (halophytes) have been grown in saline and discharge areas (Barrett-Lennard 2002) with some success. But the long-term survival in many cases is dependent on whether or not salt concentrations increase as the plants use up available water. Carefully targeted tree planting based on detailed knowledge of the local hydrogeology (especially water blocking features such as dykes) has shown the best results (Salama and Hatton 1999). However, such detailed knowledge is rarely available except from research sites or demonstration farms. In many cases, the main benefits of tree planting are seen as the establishment of habitat for specific biota and occasionally for high value forest products. In summary, broad-scale tree planting undertaken to reduce water table rises results initially at least in drying out the landscape rather than improving surface soil structure and hence does not improve the ability of the surface soil to retain more water. A significant side effect for water-limited agricultural activities is the reduction of water into streams (Zhang *et al.* 2001).

4.4.2 Tree Belts

In degraded, semiarid agricultural areas, tree belts (or strips) across hill slopes are commonly used to modify the water cycle (Fig 4.4). The main modifications to the hill slope water cycle are considered to be interception of water due to the capture of overland flow within the tree belt and hence a possible reduction to off-site water-logging and reduced recharge to salty water tables (Stirzaker *et al.* 1999, Ellis *et al.* 2006). Modifications to soil properties within ungrazed tree belts can result in increased infiltration rates due to litter build-up within

Figure 4.4 Belts of trees are used to intercept water coming from the upper slopes (photo by Tim Ellis, CSIRO).

the belts. The degree of increased infiltration rates and modified surface soil characteristics due to tree planting is, however, highly dependent on soil type (Braunack and Walker 1985). In functional terms, tree belts have been equated with naturally banded vegetation seen in arid areas, and hence "mimic" the run-on and run-off characteristics and litter terracing in these arid ecosystems (Tongway *et al.* 2001). The redistribution of overland flow and the associated nutrients either in solution or in suspended sediments can help restore some resilience and diversity to degraded agricultural systems (Hobbs and O'Connor 1990). The source–sink approach can reduce water loss and fulfils one key part of the restoration objectives—it restores some beneficial water entrainment properties in surface soils of the landscape.

4.4.3 Agroforestry and Agronomic Rotations

Integrating agroforestry with perennial pastures and crop rotations provides a mixture of deep-rooted and shallow-rooted plants (Stirzaker *et al.* 2002). It is a step toward the twin objectives of reducing lateral water movement across agricultural landscapes and at the same time maintaining productivity levels and cash flows for current agricultural systems (Dunin *et al.* 1999). The trees contribute to evapotranspiration all year round, and may also access water from the groundwater table. In areas of annual vegetation (crops, pastures) plant water uptake is very low during the nongrowing season, resulting in water leakage into the water table. Dunin (2002) and Dunin and Passioura (2006) have pointed out that many landscapes in southern Australia with original intact vegetation experience up to 5% leakage; volumes of this amount are needed to maintain year-round stream flows. However, annual crops and pastures are not to be discounted simply by their failure to reduce all leakage to the groundwater. Other benefits accrue because in Australia many perennial pastures contain a legume, such as lucerne that adds nitrogen to the system; rotations of crops and pastures are generally employed for weed control and to reduce the use of pesticides; and critically, well-managed crop-pasture rotations can build up soil organic matter.

4.4.4 Engineering Solutions

The perceived limitations of revegetation solutions using trees alone have led to more effort being put into engineering solutions. These focus on reshaping landforms, surface water removal, salty water removal, and enhanced discharge from the ground water system (McFarlane and Cox 1990, Cox and McFarlane 1995). Drainage options at local scales usually take the form of shallow drains to intercept overland flow (interceptor drains) and soil interflow, moving the water quickly off-site (Hatton 2002). The main issues with moving salty water off-site are that first, salt loads are exported further down the catchment, and second, if the water is of good quality (e.g., storm water), then it is lost from the farm. Interceptor drains have little impact on deeper ground waters moving through the landscape and can be combined with tree planting immediately down slope to try and use up this water. The impact of tree belts combined with drains can provide a localized impact on the shallow ground waters (Hatton 2002). An alternative to shallow interceptor drains is groundwater pumping combined with deep, open drains. This method is widely used at local and regional scales to remove saline waters and dispose of them into the stream networks of the region (Otto and Salama 1994). Draining and pumping can be effective in reducing the impacts of land-based salinization on urban and rural infrastructure and in maintaining an area under crops and pastures. However, moving large volumes of salty water off farm has impacts on stream biota and industries that depend on good quality water.

Natural sequence farming (NSF) is an engineering approach to manipulate the local hydrologic regime to reuse and store water within the floodplain elements of a landscape (Newell and Reynolds 2005). Small structures are used to spread stream flows out across the floodplain and to dam incised stream channels at a number of points. Small banks are also constructed at the break of slope where the surrounding hills meet the floodplain. This break reduces the velocity of water moving through the floodplain and increases aquifer water storage. Sedimentation gradually refills the incised channels and is also deposited on the floodplain, adding nutrients to the system from areas upstream. In effect, NSF is an approach to moisten the surface soils and return the floodplain to the original "chain of ponds" drainage system. The ponds are colonized by dense stands of reeds that retain nutrients and help regulate water movement off the property. To date, the NSF method has been applied extensively to a single property and there is insufficient data to quantify the impacts. However, the method does appear to improve soil organic matter, slow down deep drainage, and reduce export of salt from the property. NSF appears to include many hydrogeological and ecological principles and there is visual evidence of improved productivity, especially during drought periods. But potential applications elsewhere have several restrictions: NSF will be limited to local groundwater systems with low salt content and to landscapes that have surrounding hills to supply fresh water and sediment to the floodplain. In addition, NSF as currently implemented is in areas that previously had chains of ponds. Nevertheless, the broad principles in NSF that focus on improving soil organic matter and moistening surface soils rather than drying them out, follow the proposed end-points of restoration in the salinized landscapes outlined in 4.3.

4.4.5 Boosting Soil Organic Matter

The importance of organic matter and diverse soil biota in the restoration of salinized areas is poorly represented in the literature. Yet organic matter can become a major regulator of water movement into the subsoil. Higher surface-soil moisture levels will also improve the function of soil biota in regulating hydraulic properties and in biochemical cycling. The importance of soil microbes and fauna in regulating many ecosystem functions is well articulated by Neher (1999), Lavelle (1997), and Lavelle and Spain (2001; also see Chapter 3). Developing the means to adjust soil microbe populations to help restore hydraulic functions in old landscapes is a key research need. In production farming systems the use of mechanical methods to sow crops into stubble and reuse the stubble as mulch, have been developed over many years.

Our conclusion is that no single approach is likely to improve landscape health in all situations and in most cases a combination of approaches aimed at the belowground systems (abiotic restoration) is the most likely approach to succeed. In particular, building up the water retention capacity of the surface 50 cm of soil is essential.

4.5 A Possible Restoration Scenario

The broad objectives outlined earlier to reduce the impact or spread of salinity in southern Australia involve the initial use of broad-acre herbaceous species as rotations of crops and pastures or salt-tolerant shrubs in discharge areas to stabilize the initially degraded system. Current activities need to be coupled with soil organic matter conservation strategies (e.g., managing grazing pressures, stock exclusion, stubble retention, and composting) to improve soil moisture retention. Where terrain allows, shallow drains can harvest surface water, especially after storms. This water should have a low salt content and can be stored for reuse or to supplement environmental flows down-stream. Belts of deep-rooted perennial shrubs or legumes around the shallow drains can increase soil nitrogen, increase soil organic matter, and reduce soil movement. Planting belts of trees of varying areal extent and species composition can achieve a range of objectives including reduced run-off and sediment movement, improved surface soil properties, specialized timber production, honey production, establishing specific habitats, and wildlife corridors. On crests and upper slopes pasture species used initially to stabilize the system can be replaced with deep-rooted shrubs and trees. Species selection should take into account habitat requirements of local fauna and wildlife corridor needs. Halophytes can be established in discharge areas and will persist if salt inputs from the surrounding landscape are reduced.

The conceptual temporal (successional) stages that fulfill the above restoration scenario for a production landscape are shown in Fig. 4.5. The design is an elaboration of the "use water where it falls" concept described by McDonagh

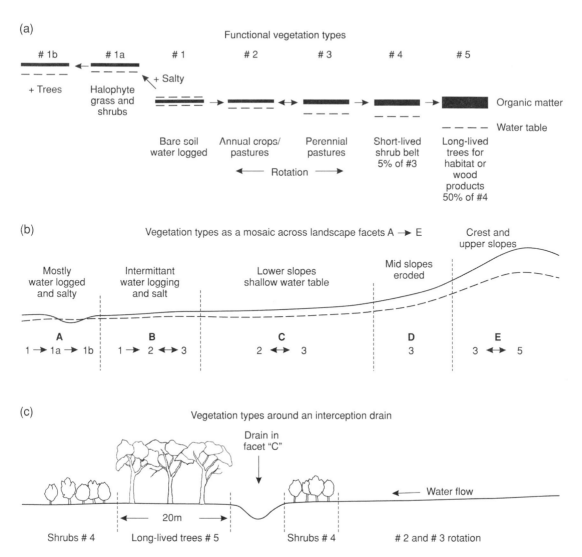

Figure 4.5 The top figure (a) shows a hypothetical five step succession to reduce water recharge. It comprises plant species selected to increase surface soil organic matter with progressively deep-rooted plants; (b) shows the vegetation types needed across the five landscape facets and (c) shows the vegetation types around an interceptor drain.

et al. (1979) for the restoration of a military training area, the source–sink model described by Ellis *et al.* 2006 and the tree planting in cropping land described by Stirzaker *et al.* (1999). The hypothetical succession (4.5a) goes from bare soil to annual grasses, perennial grasses, and forbs to shrubs to tree-dominated areas. The sequence focuses on restoring the hydrological functioning of the surface soil. Fig. 4.5a shows organic matter increasing and a gradual increase in the depth to the salty water table. The idea is to design a landscape that over time retains water, reduces leakage, and yet enables production farming to continue. The landscape facets (geomorphic or terrain units) and the successional stages are spread as a mosaic across the stylized landscape in Fig. 4.5b.

The spatial spread of "successional" stages replaces time with space and in most cases only two or three steps in the succession are needed in any one landscape facet. In a landscape with high surface run-off, the first step is to install an interceptor drain to remove excess water and store it in local dams. Perennial pastures are the precursor of shrubs and trees that are established along the drain (Fig. 4.5c). Facet A is the lowest part of the landscape and initially requires restoration with halophytes but as the water table is lowered by actions upslope and leaching of salt, true halophytes can be replaced over time by salt-tolerant trees or taller shrubs. The lower slopes (B) and (C) are the production areas and rotate between crop production and perennial grazing pastures. Steeper slopes are less suited to crop production and are placed under permanent pastures, while the steep upper slopes and crests are planted to trees as wildlife corridors and to reduce recharge from these areas.

The proportions of the different plant communities can be estimated using relatively simple and well-understood models of plant water use. Dunin (2002) has shown how the species ratio can be calculated for a farm in Western Australia. In this case, he estimated that 12% trees, 30% lucerne, and 58% annual crops would on average result in only a small leakage of water to the ground water table to maintain stream flows.

4.6 Secondary Rainforest Successions on Old Tropical Landscapes: Symptoms of Declining Site Nutrient Capital

Old, highly weathered soils cover almost three quarters of the humid tropics (Kauffman *et al*. 1998, Sanchez 1976). Many of these areas naturally support tall-stature, biologically complex rainforests where the apparent lushness of the vegetation belies the inherent infertility of the substrate (Richards 1952). It is typical in these systems on older landscapes that a high proportion (50–90%) of the total site nutrient capital (i.e., the total amount of nutrient in all pools that are potentially available to the biota) is held in the biomass (Nye 1960, Jordan 1982, Medina and Cuevas 1989) and that the vegetation has well-developed mechanisms for the acquisition and retention of nutrients (Grubb 1977, Stark and Jordan 1978, Janos 1983). Whittaker (1970) succinctly summarized the functioning of tropical rainforests on these highly weathered soils as "a relatively nutrient rich economy perched on a nutrient poor substrate."

Tropical forests on old landscapes in which most of the site nutrient capital is stored in the biomass are particularly vulnerable to high intensity anthropogenic disturbances (e.g., intensive timber extraction, burning, and clearing) and there is an extensive literature on the problematic nature of their regeneration (Sanchez 1976, Lovejoy 1985, Sim and Nykvist 1991). In these situations, not only is much of the site nutrient capital directly removed by the disturbance, but the relatively tight, biologically controlled nutrient cycles are disrupted further, exacerbating nutrient leakage from the landscape. Poor natural regeneration characterized by arrested succession and/or lack of complex structural development is a particularly common feature of these disturbed areas. In essence, disturbed tropical forests on old landscapes provide examples of where anthropogenic activities have accelerated the process of retrogressive succession.

Perhaps the most striking illustration contrasting tropical forest succession in young and old landscapes comes from work on secondary successional sequences in lower montane rainforests on the Atherton Tablelands in North Queensland (Reddell, Hopkins and Spain, unpublished). The original rainforests of the Atherton Tableland were extensively cleared for agricultural development in the early 20th century (Winter *et al.* 1987), but from the 1930s until recently, many areas were abandoned for economic reasons. By interpreting a sequence of aerial photographs of the region from the 1940s to the late 1990s, two replicated chronosequences of secondary rainforest succession on each of the two soil types (that contrasted strongly in their inherent fertility) were established. The secondary successional chronosequences were selected to include five distinctive structural/seral categories of vegetation. These categories were (1) herbland, (2) shrubland, (3) young secondary forest (approximately 18–25 years of regrowth since abandonment), (4) old secondary forest (approximately 50–70 years of regrowth since abandonment) and (5) intact rainforest that had been selectively logged but not cleared. The two soil types for which the chronosequences were identified differed markedly in their chemical and physical properties and in the structural type of rainforest community that they were capable of supporting. The contrasting soil types were (1) an old, highly weathered soil derived from Palaeozoic schists and phyllites and (2) a young, highly fertile, weakly weathered soil developed from Quaternary basaltic lavas. The two soils originally supported distinctive rainforest structural types, a simple and a complex notophyll vine forest, respectively (Tracey 1982).

Aspects of the floristics, biological productivity, rates of nutrient cycling, and nutrient acquisition strategies by plants were then compared across the stages of rainforest succession on the young and old soils, revealing major differences in the rates and trajectories of major processes and plant attributes between the contrasting sequences (Fig. 4.6).

In the case of the succession on the young soil type, measures of ecosystem function such as soil organic carbon, microbial biomass, and bulk density all increased consistently and progressively along the chronosequence and in the older secondary forests approached values in the comparable primary forest type (i.e., there was a progressive succession; Fig. 4.6). Exchangeable cations and various compositional and structural parameters, including species diversity, average canopy height, and basal area, showed the same patterns. In contrast,

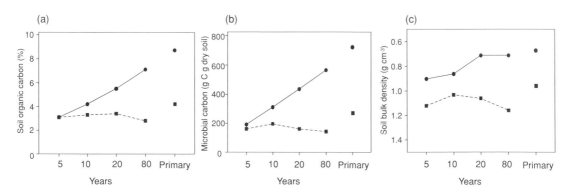

Figure 4.6 Soil organic carbon (a), microbial biomass (as measured by microbial carbon) (b), and soil bulk density (c) for rainforest successional sequences on young (solid line) and old soils (dotted line).

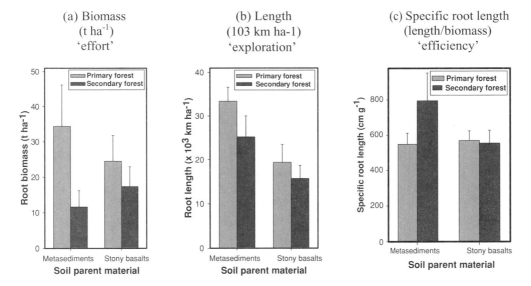

Figure 4.7 Comparison of community-level root attributes in old secondary regrowth (50–70 years) and primary rainforest on young (stony basalt) and old (metasediment) soils.

on the old soil type, although average canopy height and basal area increased along the chronosequence, most other measures (Fig. 4.6) remained largely unchanged or declined with age of development. In no cases did the values for these parameters approach those found in the matching primary forest, clearly showing a retrogressive succession.

Similarly, as succession progressed, different trends were found between the soil types in plant attributes related to nutrient acquisition (Fig. 4.7). For example, measures of root biomass, root length, and specific root length (i.e., the efficiency of roots in exploring the soil expressed as length per unit biomass) were very different between the two soil types. On the young basalt derived soil, all three measures in the oldest of the secondary forests were approaching those found in the primary forest (Fig. 4.7). In contrast, total root biomass in the oldest secondary forest on the highly weathered metasediments was still less than one-third of that in the primary forest on the same soil, while the specific root length in the secondary forest was 30% higher than in the primary forest.

The increased prevalence at the plant community level of root strategies for exploring soil as efficiently as possible during a rainforest succession on an old landscape further reinforces the importance of recognizing that natural succession in such systems is unlikely to lead back to the original plant community. Different sets of plant attributes to cope with a different system state are likely to be selected for, and these may not necessarily be inherent in the species that comprised the original community.

The above discussion has focused on nutrient issues associated with retrogressive rainforest succession in old tropical landscapes to illustrate the system changes that can occur and the associated ecosystem and plant responses. Hydrological impacts can be equally significant (Bruijnzeel 1990). For example, increases in soil bulk density and decreases in infiltration rates are common as soils age or as surface soils are eroded or removed by anthropogenic activities.

4.7 Retrogressive Rainforest Successions: What Lessons for Restoration?

Recognizing and understanding the widespread occurrence and nature of retrogressive succession on old soils in the humid tropical is fundamental to successfully and predictably restoring sustainable and functional forest cover to these areas.

Activities to restore forest cover in the humid tropics are varied in their primary purpose. These can range from habitat replacement for rare species conservation, through to tree establishment for catchment protection or for plantation forestry. Each purpose is likely to have different objectives in relation to productivity, species composition, and ecosystem function, as well are require different implementation methods suited to the purpose, budget, and local conditions. Where these activities are planned in old landscapes, there are two broad approaches that can be applied to deal with the implications of retrogressive succession. These are:

1. Alter the site conditions to ameliorate as far as possible the degraded state of the disturbed system and restore its pre-disturbance state.
2. Recognize the new system state and set realistic endpoint, composition, and productivity objectives consistent with this post-disturbance, retrogressed state.

In some instances, especially where the main impacts of disturbance have been on site nutrient capital, it may be possible to use a well-planned fertilizer strategy to recapitalize limiting nutrients into the disturbed area and return it to a potential state and vegetation type closer to the original. However, this is often likely to be expensive and resource intensive. Interestingly, the most limiting nutrient after disturbance in many old landscapes is calcium (Nykvist 1998), which is accumulated in wood and bark and has an essential role in lignin formation. Treatments to ameliorate the effects of anthropogenically accelerated retrogression on hydrological characteristics may also be technically possible, especially to increase surface infiltration, however these are unlikely to be economically feasible on the broad scale.

A less resource intensive approach that is likely to be more widely applicable is to recognize the biophysical constraints in the new post-disturbance, system state and develop endpoint objectives for composition, structure, productivity, and/or function based on this state. In many areas there may be landscape "analogues" of the new state, where natural erosion and weathering have progressed more rapidly in particular landscape locations and where existing plant communities can be used as models for species composition, structure, productivity, and function in the restoration process.

4.8 What Do We Need to Know to Restore Old Disturbed Landscapes?

Fundamental to the restoration of old disturbed landscapes is recognition that:

1. Retrogressive succession (in which total nutrient capital declines and hydrologic functioning changes as systems age) is a normal, long-term, natural process in old landscapes.

2. Major disturbances (whether natural or anthropogenic) frequently accelerate the mechanisms that underpin system retrogression, with the impact depending on (a) the state of regression in the system prior to disturbance, and (b) the nature of the disturbance event (i.e., intensity, scale, and what features of the biophysical environment were most affected).

As a consequence of the above, the biophysical environment in the post-disturbance landscape may no longer be amenable to support the original vegetation type and may be more suited to a different vegetation composition and structure that is better adapted, more sustainable, and more stable under the physicochemical and hydrological conditions that characterize the new system "state." Defining the biophysical characteristic of this new state is the key to successful restoration, but how can we do this and what are the main steps?

Although a huge diversity of vegetation structural types and compositions occur on old landscapes (reflecting the myriad of physical environments and of adaptations, attributes, and assemblages of local biota), it is possible to distill an approach to restoration based on understanding the implications of retrogressive succession down to seven practical steps. These are:

1. Determine the extent and stage of retrogression of the system prior to disturbance. This is observational ecology based, if possible, on intact areas on similar nearby landscapes. It involves the use of local knowledge, field-work, and published literature to establish the broad biophysical constraints, successional trends, and links with land-use and previous disturbance.
2. Establish the extent of the abiotic changes that have been caused by the most recent disturbance. In essence these are the abiotic constraints that will control the new state in which the system will be sustainable. In particular, characterize what have been the major changes in soil hydraulic, chemical, and biological properties. This could involve soil and terrain analysis (nutrient status, organic matter content, compaction, and infiltration) and digital elevation modeling (potential for sediment and organic matter transport)
3. Identify, where possible, areas in the surrounding landscape where similar abiotic constraints associated with natural erosion or past disturbances may be present and have resulted in a stable state that could act as "landscape analogues" for the system requiring restoration.
4. Establish what the major species were in the pre-disturbance site and in any natural landscape analogues. As far as possible understand their life cycles and disturbance responses (phenology, root strategies, seed banks, regeneration characteristics, growth rates, plasticity). This information will provide insights into the appropriate species mix, how this can feasibly be introduced and whether specific management interventions can accelerate the progress of the new system toward a stable state.
5. Identify any options for management interventions to fully or partially ameliorate the retrogression that has occurred, e.g., can critical nutrients be applied to return the site nutrient capital to the pre-disturbance state, and is this economically and technically feasible? Can aspects of hydrological function be restored by soil amendment or physical or engineering treatments?
6. Establish endpoint criteria (e.g., vegetation structure, composition, productivity, and/or function) that are compatible with the intended land use (production systems, conservation, and stabilization) for the new "state." Where

restoration of stable plant communities for conservation purposes is required try to match these as closely as possible to landscape analogue communities.

7. On the basis of points 1 to 6 above, establish the key system drivers and design the restoration program (objectives, implementation methods, costs, and monitoring protocols).

Acknowledgments: This chapter is dedicated to our friend, the late Cliff Thompson, O.A. formerly of CSIRO Division of Soils. His vast knowledge of Australian soils, the dunes at Cooloola and his pioneering work in restoring mined sites are monuments to his memory. Comments by Tim Ellis, David Mackenzie, Karel Prach, Simon Veitch, and Lawrence Walker greatly improved this chapter.

References

Allison, G. B., and Hughes, M. W. 1983. The use of natural tracers as indicators of soil water movement in temperate semi-arid regions. *Journal of Hydrology* 60:157–173.

Barrett-Lennard, E. G. 2002. Restoration of saline land through revegetation. *Agricultural Water Management* 53:213–226.

Bell, R. W., Schofield, N. J., Loh, I. C., and Vari, M. A. 1990. Groundwater response to reforestation in the Darling Ranges in southwestern Australia. *Journal of Hydrology* 115:297–317.

Blackburn, G. 1976. Salinity cycles in Australia, with special reference to the Murray Basin. *Transactions of the 10th Congress of the International Soil Science Society*, Moscow, Vol. IV, pp. 27–33.

Braunack, M. V., and Walker, J. 1985. Recovery of some surface soil properties of ecological interest after removing sheep grazing in a semi-arid woodland. *Australian Journal of Ecology* 10:451–460.

Bruijnzeel, L. A. 1990. *Hydrology of Moist Tropical Forests and Effects of Conversion: A State of Knowledge Report*. Paris: UNESCO.

Cowles, H. C. 1899. The ecological relations of the vegetation on the sand dunes of Lake Michigan. *Botanical Gazette* 27:95–117, 167–202, 281–308, 361–369.

Cox, J. W., and McFarlane, D. J. 1995. The causes of waterlogging in shallow soils and their drainage in southwestern Australia. *Journal of Hydrology* 167:175–194.

Davis, W. M. 1899. The geographical cycle. *Geographical Journal* 14:481–504.

Dunin, F. X., Williams, J, Verburg, K., and Keating, B. A. 1999. Can agricultural management emulate natural ecosystems in recharge control in south eastern Australia? *Agroforestry Systems* 45:343–364.

Dunin, F. X. 2002. Integrating agroforestry and perennial pastures to mitigate water logging and secondary salinisation. *Agricultural Water Management* 53:259–270.

Dunin, F. X., and Passioura, J. 2006. Prologue: Amending agricultural water use to maintain production while affording environmental protection through control of outflow. *Australian Journal of Agricultural Research* 57:251–255.

Ellis, T. W., Leguedois, S., Hairsine, P. B., and Tongway, D. J. 2006. Capture of overland flow by a tree belt on a pastured hillslope. *Australian Journal of Soil Research* 44:117–125.

Ellis, T. W., Hatton, T. J., and Nuberg, I. 2005. An ecological optimality approach for predicting deep drainage from tree belts of alley farms in water-limited environments. *Agricultural Water Management* 75:92–116.

Fitzpatrick, R.W., Fritsch, E., and Self, P. G. 1996. Interpretation of soil features produced by ancient and modern processes in degraded landscapes: V. Development of saline sulfidic features in non-tidal seepage areas. *Geoderma* 69:1–29.

George, R. J., McFarlane, D. J., and Nulsen, R. A. 1997. Salinity threatens the viability of agriculture and ecosystems in Western Australia. *Hydrogeology Journal* 5:6–21.

George, R. J., Nulsen, R. A., Ferdowsian, R., and Raper, G. P. 1999. Interactions between trees and groundwaters in recharge and discharge areas – A survey of Western Australian Sites. *Agricultural Water Management* 39:91–113.

Greenwood, E. A. N., Milligan, A., Biddiscombe, E., Rogers, A. L., Beresford, J. D., and Watson, G. D. 1992. Hydrologic and salinity changes associated with tree plantations in a saline agricultural catchment in southwestern Australia. *Agricultural Water Management* 22:307–323.

Grubb, P. J. 1977. Control of forest growth and distribution on wet tropical mountains: With special reference to mineral nutrition. *Annual Review of Ecology and Systematics* 8:83–107.

Gunn, R. H., and Richardson, D. P. 1979. The nature and possible origins of soluble salts in deeply weathered landscapes of eastern Australia. *Australian Journal of Soil Research* 17:197–215.

Hatton, T. J. 2002. Engineering our way through Australia's salinity problem. CSIRO Land and Water Technical Report 4/02.

Hatton, T. J. and Nulsen, R. A. 1999. Towards achieving functional mimicry with respect to water cycling in southern Australian agriculture. *Agroforestry Systems* 45:203–214.

Hedin, L. O., Vitousek, P. M., and Matson, P. A. 2003. Nutrient losses over four million years of tropical forest development. *Ecology* 84:2231–2255.

Herczeg, A. L., Dogramaci, S. S., and Leaney, F. W. J. 2001. Origin of dissolved salts in a large, semi-arid groundwater system: Murray basin, Australia. *Marine and Freshwater Research* 52:41–52.

Hobbs, R. J., and O'Connor, M. H. 1990. Designing mimics from incomplete data sets: Salmon gum woodland and heath ecosystems in South West Australia. In: Agriculture as a Mimic of Natural Ecosystems, pp. 432–436. E. C. Lefroy, R. J. Hobbs, and J. S. Pate (eds.). *Agroforestry Systems*, Special Issue No. 54(1–3).

Iversen, J. 1964. Retrogressive vegetational succession in the Post-glacial. *Journal of Ecology* 52(Suppl.):59–70.

Janos, D. P. 1983. Tropical mycorrhizas, nutrient cycles and plant growth. In: *Tropical Rain Forest: Ecology and Management*. S. L. Sutton, T. C. Whitmore, and A. C. Chadwick (eds.). Oxford: Blackwell, pp. 327–345.

Jordan, C. F. 1982. The nutrient balance of an Amazonian rainforest. *Ecology* 61:14–18.

Kauffman, S., Sombroek, W., and Mantel, S. 1998. Soils of rainforest: Characterization and major constraints of dominant forest soils in the humid tropics. In: *Soils of Tropical Forest Ecosystems*. A. Schulte and D. Ruhiyat (eds.). Berlin: Springer, pp. 9–20.

Lamont, B. B. 1981. Specialized roots of non-symbiotic origin in heathlands. In: *Ecosystems of the World. 9B. Heathlands and Related Shrublands*. R. L. Specht (ed.). Amsterdam: Elsevier, pp. 183–195.

Lavelle, P. 1997. Faunal activities and soil processes; adaptive strategies that determine ecosystem function. *Advances in Ecological Research* 27:93–132.

Lavelle, P., and Spain, A. V. 2001. *Soil Ecology*. Dodrecht: Kluwer.

Lovejoy, T.E. 1985. Rehabilitation of Degraded Tropical Lands. IUCN Commission on Ecology, Occasional Paper No. 5.

McDonagh, J. F., Walker, J., and Mitchell, A. 1979. Rehabilitation and management of an army training area. *Landscape Planning* 6:375–390.

McFarlane, D. J., and Cox, J. W. 1990. Seepage interceptor drains for reducing waterlogging and salinity on clay flats. *Journal of Agriculture Western Australia*. 31:70–73.

McFarlane, D. J., and Williamson, D. R. 2002. An overview of waterlogging and salinisation in southwestern Australia as related to 'Ucarro' experimental area. *Agricultural Water Management* 53:5–29.

Medina, E., and Cuevas, E. 1989. Patterns of nutrient accumulation and release in Amazonian rainforests of the Upper Rio Negro. In: *Mineral Nutrients in Tropical Forest and Savannah Ecosystem*. J. Proctor (ed.). Oxford: Blackwell, pp. 217–240.

Neher, D. A. 1999. Soil community composition in ecosystem processes: Comparing agricultural ecosystems with natural ecosystems. In: Agriculture as a mimic of natural ecosystems (pp. 159–185). E. C. Lefroy, R. J. Hobbs, and J.S. Pate (eds.). *Agroforestry Systems*, Special Issue No. 54(1–3).

Newell, P., and Reynolds, G. 2005. Natural sequence farming: Principles and applications. http://www.nsfarming.com/

NLWRA. 2001. *National Land and Water Resources Audit. Australian Dryland Salinity Assessment 2000. Extent, Impacts, Processes, Monitoring and Management Options*. Canberra: Land and Water Australia.

Northcote, K. H., and Skene, J. K. M. 1972. *Australian Soils with Saline and Sodic Properties*. CSIRO, Collingwood, Australia. Soil Publication No. 27.

Nye, P. H. 1960. Organic matter and nutrient cycles under moist tropical forest. *Plant and Soil* 13:333–346.

Nykvist, N. 1998. Logging can cause a serious lack of calcium in tropical rainforest ecosystems: An example from Sabah, Malaysia. In: *Soils of Tropical Forest Ecosystems*. A. Schulte and D. Ruhiyat (eds.). Berlin: Springer, pp. 87–91.

Otto, C., and Salama, R. B. 1994. Linked enhanced discharge-evaporative disposal basins. In: *Groundwater – Drought, Pollution and Management*. R. Reeve and J. Watts (eds.). Rotterdam: A.A. Balkema, pp. 35–44.

Pate, J. S. 1994. The mychorrizal association: Just one of many nutrient acquiring specialisations in natural ecosystems. *Plant and Soil* 159:1–10.

Richards, P. W. 1952. *The Tropical Rain Forest*. Cambridge: Cambridge University Press.

Salama, R. B., and Hatton, T. J. 1999. Predicting land use impacts on regional scale groundwater recharge and saline discharge. *Journal of Environmental Quality* 28:446–460.

Sanchez, P. A. 1976. *Properties and Management of Soils in the Tropics*. New York: Wiley.

Silberstein, R. P., Vertessy, R. A., Morris, J., and Feikma, P. M. 1999. Modelling the effects of soil moisture and solute conditions on long term tree growth and water use: A case study from the Shepparton irrigation area, Australia. *Agricultural Water Management* 39:283–315.

Sim, B. L., and Nykvist, N. 1991. Impact of forest harvesting and replanting. *Journal of Tropical Forest Science* 3:251–284.

Stark, N. 1978. Man, tropical forests and the biological life of a soil. *Biotropica* 10:1–10.

Stark, N., and Jordan C. F. 1978. Nutrient retention by the root mat of an Amazonian rainforest. *Ecology* 59:434–437.

Stirzaker, R. J., Cook, F. J., and Knight, J. H. 1999. Where to plant trees on cropping land for control of dryland salinity: Some approximate solutions. *Agricultural Water Management* 39:115–133.

Stirzaker, R. J., Lefroy, E. C., and Ellis, T. W. 2002. An index for quantifying the trade-off between drainage and productivity in tree-crop mixtures. *Agricultural Water Management* 53:187–199.

Summerell, G. K., Dowling, T. I., Richardson, D. P., Walker, J., and Lees, B. 2000. Modelling current parna distribution in a local area. *Australian Journal of Soil Research* 38:867–878.

Thompson, C. H. 1981. Podzol chronosequences on coastal dunes in eastern Australia. *Nature* 91:59–61.

Thompson, C. H. 1983. Development and weathering of large parabolic dune systems along the sub-tropical coast of eastern Australia. *Zeits Geomorphological Supplement Band* 45:205–225.

Thompson, C. H. 1992. Genesis of podzols on coastal dunes in southern Queensland. Field relationships and profile morphology. *Australian Journal of Soil Research* 30:593–613.

Thorburn, P. J. 1997. Land management impacts on evaporation from shallow, saline water tables. In: *Sub-Surface Hydrological Responses in Land Cover and Land-Use Changes.* M. Taniguchi (ed.). Boston: Kluwer, pp. 21–34.

Thorburn, P. J., Walker, G. R., and Jolly, I. D. 1995. Uptake of saline groundwater by plants: An analytical model for semi-arid and arid areas. *Plant and Soil* 175:1–11.

Tongway, D. J., Valentin, C., and Seghieri, J. 2001. *Banded Vegetation Patterning in Arid and Semi-Arid Environments—Ecological Processes and Consequences for Management.* New York: Springer. Ecological Studies Series No. 149.

Tracey J. G. 1982. *The Vegetation of the Humid Tropical Region of North Queensland.* Melbourne: CSIRO.

Vitousek, P. M., and Reiners, W. A. 1975. Ecosystem succession and nutrient retention: A hypothesis. *BioScience* 25:376–381.

Walker, G. R., Zhang, L., Ellis, T. W., Hatton, T. J., and Petheram, C. 2002. Estimating the impact of changed land-use on recharge: Review of modelling and other approaches as appropriate for dryland salinity management. *Hydrogeology Journal* 10:68–90.

Walker, J., Thompson, C. H., Fergus, I. F., and Tunstall, B. R. 1981. Plant succession and soil development in coastal sand dunes of subtropical eastern Australia. In: *Forest Succession, Concepts and Application.* D. C. West, H. H. Shugart, and D. B. Botkin (eds.). New York: Springer, pp. 107–131.

Walker, J., Thompson, C. H., and Lacey, C. J. 1987. Morphological differences in lignotubers of *Eucalyptus intermedia* R.T. Bak. and *E. signata* F. Muell. associated with different stages of podzol development on coastal dunes, Cooloola, Queensland. *Australian Journal of Botany* 35:301–311.

Walker, J., and Nicholl, C. 1996. Assessing tree planting for land degradation management: Waterlogging and dryland salinity. In: *Proceedings of the 23rd Hydrology and WaterResources Symposium, Hobart,* 1996.

Walker, J., Thompson, C. H., Reddell, P. and Rapport, D. J. 2001. The importance of landscape age in influencing landscape health. *Ecosystem Health* 7:7–14.

Walker, T. W., and Syers, J. K. 1976. The fate of phosphorus during pedogenesis. *Geoderma* 15:1–19.

Wardle, D. A., Walker, L. R., and Bardgett, R. D. 2004. Ecosystem properties and forest decline in contrasting long-term chronosequences. *Science* 305:509–513.

Warming, E. 1895. *Plantesamfund: Grunträk af den Ökologiska Plantegeografi.* Copenhagen: Philipsen.

Whittaker, R. H. 1970. *Communities and Ecosystems.* London: MacMillan.

Williams, B. G., Walker, J. and Anderson, J. 2006. Spatial variability of regolith leaching and salinity in relation to whole farm planning. *Australian Journal of Experimental Agriculture* 46:1271–1277.

Williams, B. G., Walker, J., and Tane, H. 2001. Drier landscapes and rising water tables. *Natural Resource Management* 4:10–18.

Winter, J. W., Bell, F. C., Pahl, L. I., and Atherton R. G. 1987. Rainforest clear-felling in northeastern Australia. *Proceedings of the Royal Society of Queensland* 98:41–57.

Zhang, L., and Dawes, W .R. 1998. WAVES: An integrated energy and water balance model (Published as a website http://www.clw.csiro.au/products/waves/). CSIRO Land and Water Technical Report 31/98.

Zhang, L., Dawes, W. R., and Walker, G. R. 2001. Response of mean annual evapotranspiration to vegetation changes at catchment scales. *Water Resources Research* 37:701–708.

5

Succession and Restoration of Drained Fens: Perspectives from Northwestern Europe

Joachim Schrautzer, Andreas Rinker, Kai Jensen, Felix Müller, Peter Schwartze, and Klaus Dierßen

Key Points

1. Changes of ecosystem features are described for several stages of retrogressive succession due to land-use intensification and secondary progressive succession due to abandonment in European fens.
2. Agricultural intensification causes a loss of ecosystem features in fens, which in turn reduces the capacity of fens to provide multiple ecosystem services.
3. Abandonment will not always lead to the development of the desired successional stages, so technical manipulation such as rewetting, grazing, or mowing may be needed.

5.1 Introduction

Fens are peat-forming wetlands that receive nutrients from upslope sources through drainage from surrounding mineral soils and from groundwater movement (EPA 2006). Fens are dominated by graminoids (grasses, sedges, and reeds), which distinguishes them from bogs (the other major type of mire or peat-forming ecosystem) that are dominated by mosses. Globally, fens are a widespread type of wetland (Mitsch and Gosselink 2000, Fraser and Keddy 2005), covering large areas of the Holarctic boreal zone (North America, Scandinavia, eastern Europe, and western Siberia). They are also regionally abundant in tropical Southeast Asia, temperate South America, and New Zealand at high elevations. Various kinds of fens receive different amounts of precipitation and different proportions of surface water and drainage water from their catchment areas. Water tables in pristine fens fluctuate near the peat surface. Due to these abiotic conditions, fens supply several important ecological services, including the ability to retain or convert nutrients. Moreover, they offer habitats for rare, hygrophilous plants and animals.

In most countries of northwestern Europe (e.g., The Netherlands, the United Kingdom, Denmark, Germany) more than 90% of fen ecosystems have been transformed into meadows and pastures during the last several decades (Rosenthal *et al.* 1998, Joosten and Couwenberg 2001). As a consequence, pristine fens belong to one of the most threatened ecosystems of these countries

Table 5.1 Species composition of fen plant communities in northwestern Europe.

Plant community	Common plant species
Alnion glutinosae (wet alder carrs)	*Alnus glutinosa, Carex elongata, Carex elata, Thelypteris palustris*
Alnion glutinosae (dry alder carrs)	*Alnus glutinosa, Urtica dioica, Galium aparine, Poa trivialis*
Caricion elatae (tall sedge reeds)	*Carex acutiformis, Carex acuta, Rumex hydrolapathum, Cicuta virosa*
Scheuchzerio-Caricetea (small sedge reeds)	*Carex rostrata, Carex nigra, Viola palustris, Menyanthes trifoliata*
Molinietalia (wet meadows)	*Caltha palustris, Silene flos-cuculi, Lotus uliginosus, Myosotis scorpioides*
Lolio-Potentillion (wet pastures)	*Agrostis stolonifera, Glyceria fluitans, Alopecurus geniculatus, Ranunculus repens, Juncus articulatus, Lolium perenne*
Arrhenatheretalia (mesic grasslands)	*Festuca pratensis, Rumex acetosa, Poa pratensis, Taraxacum officinale*

(Pfadenhauer and Grootjans 1999; see Chapter 6). Widespread fen degradation has led to the loss of many ecological services resulting in high carbon losses to the atmosphere and nutrient additions to ground and surface waters. Furthermore, intensive land uses have caused a decrease in the number of characteristic and rare fen species.

Several studies have shown that because of increasing lack of interest to use fens for agricultural purposes farmers began to abandon them in the 1970s and more fens continue to be abandoned to this day. Abandonment involves cessation of management inputs and active land use. A major conservation issue in northwestern Europe arises from the abandonment of seminatural fen ecosystems and the loss of species found in *Scheuchzerio-Caricetea* (small sedge reeds) and *Molinietalia* (wet meadows) plant communities (Table 5.1). The natural habitats of these species are border areas of bogs, riparian zones of oligotrophic and mesotrophic lakes, oligotrophic discharge fens, and undisturbed riverbanks, and these habitats are almost extinct in northwestern Europe.

Recently, action plans for mire conservation and restoration based on the "Guidelines for Global Action on Peatlands" (Ramsar Convention 2002) have been developed worldwide (Bragg and Lindsay 2003, GEC 2003, Wetlands International Russia 2003). Due to the range of sociocultural conditions and fen characteristics in the different countries involved, strategies for these action plans differ. In countries such as Russia with a high proportion of undisturbed fens and bogs, protection of these areas has a high priority. By way of contrast, in other countries such as Germany with a high proportion of degenerated mires, the focus is on restoration activities.

In Germany and elsewhere in northwestern Europe, management is aimed at restoring degenerated fens and can be divided into two types. The first approach pursues the maintenance and development of species-rich, weakly degraded fens by mowing or grazing without fertilization in combination with moderate rewetting. The second aims to reestablish high water tables with periodic flooding to induce a succession that restores the nutrient retention capacity of the original fen.

Here we outline key ecosystem features (functional variables of the water, nitrogen and carbon budget, species density and composition, and seed longevity) operating in intact and degraded fens. The variables have been selected on the basis of theoretical considerations to represent the functionality of landscape units and elements of environmental thermodynamics (Joergensen 2001). This selection is achieved by coupling gradient theory (Müller 1998) and orientor theory (Müller and Fath 1998). The result is a small set of indicators capable of representing ecosystem states and ecosystem integrity as a whole, i.e., focusing on the potential to develop self-organized processes (Müller *et al.* 2006). These characteristics are represented by orientors, that is, ecosystem variables that should increase throughout an untreated succession following abandonment (Müller and Joergensen 2000). Changes in the indicators associated with fen restoration measures (rewetting, grazing, and mowing without fertilization) or abandonment are modeled using a data set from our long-term experimental sites to parameterize the model. The Water and Substance Simulation Model WASMOD of Reiche (1994) was used for the simulations. Simulation modeling is used because data concerning the long-term changes in the process drivers of the water, carbon, and nutrient budgets are scarce. We use our own data to illustrate the establishment of species on fen sites, and what happens to species richness on these sites given cattle grazing and mowing. Finally, we include data concerning the longevity of seeds in the soils to assess the potential for reintroduction of target species in degraded fens. We conclude with a discussion of factors limiting the success of these restoration measures and the broader implications for the restoration of fens elsewhere in the region.

5.2 Methods, Concepts, and Data Sets

5.2.1 The Basic Set of Ecosystem Indicators

The indicators chosen for this study have been compiled to represent ecosystem states as holistic entities. The variables refer to the following components of ecosystems (see Table 5.2):

Ecosystem structures: Following the orientor approach for successional phases in undisturbed ecosystems, it can be hypothesized that the number of species increases through time, and that abiotic mechanisms become more complex. These changes are accompanied by increased ecosystem heterogeneity and complexity.

Ecosystem functions: Ecosystem processes can follow orientor behavior during succession. Due to increasing numbers of structural elements, the processes

Table 5.2 Indicators selected to represent basic ecosystem components.

Ecosystem component	Indicator(S)
Biotic structures	Number of plant species
Energy budgets, exergy capture	Net primary production (NPP)
Energy budgets, entropy production	Microbial soil respiration (MSR)
Energy budgets—metabolic efficiency	NPP/soil respiration
Hydrological budgets—biotic water flows	NPP/transpiration
Chemical budgets—nutrient loss	Net nitrogen mineralization (NNM)
	Nitrate leaching
	Denitrification
Chemical budgets, storage capacity	Nitrogen balance
	Carbon balance

of the energy, water, and matter budgets become more complex, the significance of storage grows, and consequently the residence times of inputs increase. Due to the high degree of mutual adaptation, the efficiencies of single transfer reactions rise, cycling is optimized, and thus losses of matter are reduced. The correlated ecosystem functions are usually investigated within three classes of processes:

– *Ecosystem energy balance*: Exergy capture (uptake of usable energy) and exergy storage (biomass, organic matter, and information) regularly increase during succession (Schneider and Kay 1994, Joergensen 2001). The total system throughput increases (Odum *et al.* 2000) and the energy demand for maintenance and respiration also increases (Svirezhev and Steinborn 2001).

– *Ecosystem water balance*: As terrestrial ecosystems and landscapes develop without disturbance, more and more structural elements have to be supplied with water. Thus, water flows through vegetation compartments show a typical orientor behavior (optimization of biotic water flows, see Kutsch *et al.* 1998 and 2001). These fluxes are prerequisites for all cycling activities in terrestrial ecosystems. In this study, the hydrological features have been related to ecosystem productivity, hence representing a water-based efficiency measure.

– *Ecosystem matter balance*: During undisturbed ecosystem succession, imported nutrients are transferred throughout the biotic community with increased partitioning into more structures. Therefore, the biological nutrient fractions increase as well as the abiotic carbon and nutrient storages; the cycling rate also increases and efficiencies improve.

5.2.2 Data Sets

5.2.2.1 Successional Models

To describe retrogressive successional changes during land-use intensification we use a model based on results of repeated vegetation mappings in fen areas of northern Germany (Schrautzer 1988). Secondary succession after abandonment of fens uses the sequence described in the model of Jensen and Schrautzer (1999) that uses structural characteristics of the vegetation to define developmental stages. Most of the developmental stages have been studied on permanent plots or by repeated vegetation mapping. To assess the effect of rewetting on structure and processes of degenerated fens, we construct a successional sere that starts with intensively used wet pastures (*Lolio-Potentillion*), proceeds in time to eutrophic communities dominated by tall sedges and reeds (henceforth "tall sedge reeds," *Caricion elatae*), and ends in the long run with eutrophic wet alder carrs (*Alnion glutinosae*). The retrogressive succession from wet alder carrs to wet pastures due to land-use intensification, and the succession from wet meadows to dry alder carrs following abandonment are shown in Fig. 5.1.

5.2.2.2 Site Factors and Vegetation Parameters

To characterize the site factors and productivities of the successional stages we used data from Schrautzer (2004), Schrautzer and Jensen (2006), and unpublished data. LAI (leaf area index) data were obtained from Schieferstein (1997), Trepel (2000), and Kutsch *et al.* (2000). Species richness (vascular plants) of

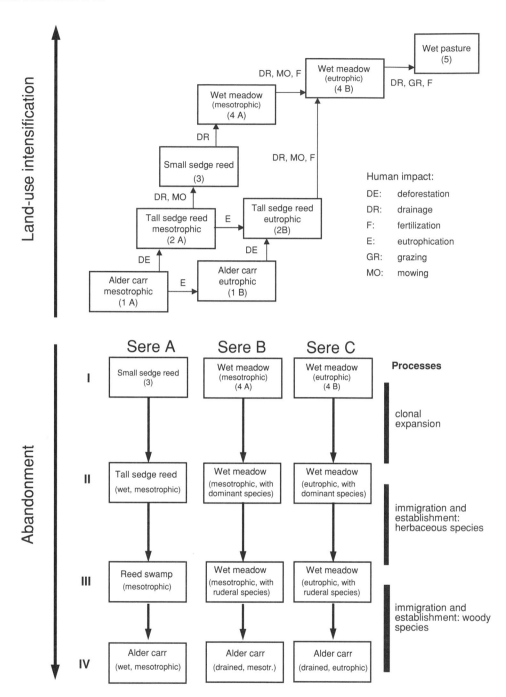

Figure 5.1 Models describing stages 1A to 5 in secondary retrogressive succession (land-use intensification), and stages I to IV for secondary progressive (abandonment) succession on seres A, B, and C on fen sites. The initial stages of the abandonment seres differ in drainage and land-use intensity. The phytosociological units are: *Alnion glutinosae* (alder carrs), *Magnocaricion elatae* (tall sedge reeds), *Scheuchzerio-Caricetea* (small sedge reeds), *Molinietalia* (wet meadows), *Lolio-Potentillion* (wet pastures).

successional stages was obtained from a data set of about 3100 releves from fen sites in Schleswig-Holstein (northernmost federal state of Germany). The data of Schrautzer and Jensen (2006) were used to develop a relationship between light availability (PAR), standing crop, and the number of small-growing species (mean height ≤30 cm) for different successional stages at the fen sites. PAR (photosynthetically active radiation) was expressed as relative irradiance (RI), which characterizes the light intensity within the stand relative to that existing above the canopy.

The relationships between the mass of seeds planted at litter depths of 3 and 8 cm and the establishment of wet grassland species originate from Jensen and Gutekunst (2003). These authors investigated 30 species from fen sites belonging to *Scheuchzerio-Caricetea*, *Molinietalia*, and *Arrhenatheretalia* (mesic grassland) plant communities.

In ecosystems where plant species composition changes during degradation, knowledge about the longevity of seed in soils is of particular importance to assess their biotic development potentials after the introduction of management measures. To determine seed longevity of fen species we used an indirect classification scheme developed by Thompson *et al.* (1997) based on extensive seed bank data from northern Germany and other European studies and our own data from seed burial in glasshouse experiments. The analysis integrates data of a seed bank database and additional results of seed bank investigations in Germany. The classification rules of Thompson *et al.* (1997) are based on (a) presence or absence in current vegetation and seed bank, (b) depth distribution of seeds in the soil, and (c) the period since the last record of a species in the current vegetation. The "Longevity-Index (LI)" of the species was calculated using the method of Bekker *et al.* (1998a). Longevity-Index is defined as the ratio of the number of short-term, persistent seed bank records (seeds viable for 1–4 years) and long-term persistent records (seeds viable for >4 years) to the sum of transient (seeds viable for <1 year) seed bank records.

5.2.2.3 Field Experiments

Species richness trends for the restoration treatments of grazing, and mowing without fertilization were analyzed using the results from 15 field experiments carried out in northwestern Europe during the last few decades. In addition, data for vegetation trends in degraded fens after introduction of different mowing regimes came from the Biological Station of Steinfurt and data for grazing regimes from the Research project "The Eidertal pasture landscape," German Research Ministry. The permanent mowing treatment plots (16 m^2) are located in the northern part of the German federal state of Northrhine-Westfalia (Schwartze 1992). A broadscale cattle grazing trial was established in 1999 in a 400 ha area in the valley of the river Eider located in the eastern part of the German regional state of Schleswig-Holstein. Here, we present results from three pastures with different vegetation and previous land-use histories. Pasture sizes vary between 30 and 40 ha. Fen area amounts to approximately 50% of the landscape. Stock density that replaced previous land-uses was 1.5 cattle/ha and grazing duration was mid-May through mid-October. The restoration history of these experiments was evaluated using changes in species richness and changes in target species (those species that characterize species-rich, weakly degraded fen communities: *Scheuchzerio-Caricetea*, *Molinietalia*).

Table 5.3 Site factors and vegetation parameters of successional (retrogressive) stages of the land-use intensification sere. For ecosystem types and codes, see Fig. 5.1. Values (means) were used to parameterize the simulation model (WASMOD) n.i. = not investigated.

Site factors	1A	1B	2A	2B	3	4A	4B	5
Ecosystem types								
Soil parameter (0–20cm):								
C/N-ratio	20	14	23	17	27	18	14	12
pH (H_2O)	3.9	5.7	5.6	5.5	5.0	5.3	5.5	5.9
Org. matter (%)	81	65	66	49	74	48	31	42
Bulk density (g cm^{-3})	0.15	0.25	0.2	0.3	0.2	0.4	0.5	0.7
Hydrological parameter:								
Mean GW-table (cm) a^{-1}	−4	−6	2	−0.5	−4.5	−14	−22	−35
Mean GW-amplitude a^{-1}	17	22	13	23	14	42	52	69
Mean flooding duration (%) a^{-1}	13	11	47	43	16	3	4	12
Vegetation parameter:								
Max. standing crop (g m^{-2})	n.i.	n.i.	509	765	105	289	437	550
Max. LAI	6	6	4	5	2.5	3.5	4.5	4
Land use:								
Fertilization (kg N ha^{-1} a^{-1})	0	0	0	0	0	0	80	140
Mowing	—	—	—	—	1 cut	1 cut	2 cuts	1 cut
Grazing	—	—	—	—	—	—	—	2–3 cattle ha^{-1}

5.2.3 Characterization of the Successional Seres

The successional sequences in the fens of northwestern Europe following land-use intensification (retrogressive succession) or abandonment (progressive succession) are presented in Fig. 5.1. The site factors and vegetation parameters associated with these seres are given in Table 5.3 (land-use intensification) and Table 5.4 (abandonment). The number of sites used in the tabulations varied from 8 to 80 (mean = 30).

5.2.3.1 Land-Use Intensification Sere

Human use of undrained alder carrs is usually low, because the mean groundwater tables at sites with these plant communities vary between 4 and 6 cm

Table 5.4 Site factors and vegetation parameters of successional (progressive) stages of the abandonment seres. Values (means) were used to parameterize the simulation model (WASMOD) n.i. = not investigated.

Sere:	A	A	A	A	B	B	B	B	C	C	C	C
Successional stage:	I	II	III	IV	I	II	III	IV	I	II	III	IV
Soil parameter (0–20cm):												
C/N-ratio	27	23	22	20	18	20	17	16	14	15	14	14
pH (H_2O)	5.0	5.6	5.5	3.9	5.3	5.4	5.2	3.5	5.5	6.1	5.8	4.8
Org. matter (%)	74	66	63	81	48	44	43	53	31	40	25	48
Bulk density (g cm^{-3})	0.2	0.2	0.2	0.15	0.4	0.4	0.4	0.5	0.5	0.5	0.5	0.6
Hydrological parameter:												
Mean GW-table (cm) a^{-1}	−4.5	2	1	−4	−14	−7	−10	−39	−22	−10	−14	−22
Mean GW-amplitude a^{-1}	14	13	15	17	42	17	20	57	52	23	31	41
Mean flooding duration (%) a^{-1}	16	47	37	13	3	20	3	0	4	11	4	0
Vegetation parameter:												
Max. standing crop (g m^{-2})	105	509	771	n.i.	289	411	505	n.i.	437	611	853	n.i.
Max. LAI	2.5	3.5	6	6	3.5	4	6	6	4.5	5	7	6

Figure 5.2 Undrained eutrophic alder carr with *Alnus glutinosa* and *Carex acutiformis* in Schaalsee, southeastern Schleswig-Holstein, Germany.

above the surface (Table 5.3). Alder carrs are subdivided into a unit dominated by mesotrophic species of the *Scheuchzerio-Caricetea* (vegetation type 1 A, Fig. 5.1) and a eutrophic unit without these species (vegetation type 1 B; Fig. 5.2). The C/N-ratio of the peat is higher at sites of the mesotrophic carr than at those of the eutrophic carr. After deforestation, alder carrs develop to tall sedge reeds (2 A, B). In most cases, the dominant sedges were present in the previous woody stages. Higher groundwater tables and longer flooding periods occur at sites of tall sedge reeds compared with sites of alder carrs, indicating the higher transpiration rate of the alders compared with the sedges. Moderate drainage and long-term mowing or grazing resulted in the development of small sedge reeds (3). A high C/N-ratio of the peat of these sites indicates low nitrogen availability. Standing crop and the LAI of small sedge reeds are much lower than the respective values for tall sedge reeds (Table 5.3). Increased drainage transforms small sedge reeds into mesotrophic wet meadows (4 A), whereas eutrophic wet meadows (4 B) develop after drainage, mowing, and fertilization. A different trajectory for small sedge reeds to wet pastures (5) results from increased drainage, higher fertilization, and grazing with high stock densities (2–3 cattle per ha). The main differences between wet meadows and wet pastures are that the groundwater tables are lower but the flooding duration is higher in wet pastures. The latter results from high compaction of the 0–20 cm peat horizon (bulk density 0.7 g cm^{-3}, Table 5.3). This compaction causes stagnant water after heavy rainfall (Schrautzer *et al.* 1996). Secondary succession from small sedge reeds via wet meadows to wet pastures resulted in an increase of standing crop and LAI of the vegetation (Table 5.3).

5.2.3.2 Abandonment Seres
The four progressive successional stages shown in Fig. 5.1 following abandonment are for seral types 3 (small sedge reed), labeled Sere A, 4A (mesotrophic

Figure 5.3 Mowed foreground and abandoned background (successional stage III with dominant *Urtica dioica*) wet meadow (*Molinietalia*) in Schmalensee, central Schleswig-Holstein, Germany.

wet meadow), labeled Sere B, and for 4B (eutrophic wet meadow), labeled Sere C. Stage 1 of each of the abandoned sites had a land-use history of cattle grazing and mowing for hay.

All stage II sites are wet sites and are characterized by an increase in tall clonal species such as *Calamagrostis canescens, Carex acutiformis, Carex acuta, Filipendula ulmaria, Glyceria maxima, Juncus acutiflorus, Juncus subnodulosus,* or *Phalaris arundinacea*. These species were already present in stage I, but became dominant only after abandonment.

In sere (A), stage III develops from stage II through immigration and establishment of herbaceous species (*Phragmites australis*). In seres (B) and (C), ruderal species such as *Urtica dioica, Galium aparine,* and *Galeopsis tetrahit* agg. increase in dominance (Fig. 5.3). These species seldom occur in vegetation of stage I. Stage IV in all seres results from the immigration and establishment of woody species (*Alnus glutinosa, Betula pubescens*).

Site factors and vegetation parameters in all investigated seres differed in a characteristic way (Table 5.4). Groundwater tables increase and groundwater amplitudes decrease from stages I to II and vice versa from stages III to IV. Standing crop and LAI increase significantly from stages I to III (cf. Schrautzer and Jensen 2006). The lowest standing crop (105 g m^{-2}) and LAI (2.5) values were detected in stands of stage I of sere A.

Even though the vegetation trends for the abandonment seres have been shown in several successional studies, it is impossible to predict the time when one successional stage passes over to the next. This is due to differences in surrounding vegetation and land-use history resulting in a specific seed bank composition and potentials for the immigration of new species. However, data from the Biological Station of Steinfurt (unpublished) show the transition from stage I up to stage III occurs more quickly at eutrophic sites than at mesotrophic ones. The expansion of clonal species (in this case *Juncus acutiflorus*) starts earlier at the eutrophic sites.

5.2.4 Water and Nutrient Budgets

The indicators in Table 5.2, were simulated with the "Water and Substance Simulation Model" WASMOD (Reiche 1994). This model incorporates the processes of the water, nitrogen, and carbon budgets on different temporal and spatial scales. In the present study, we used the data in Tables 5.3 and 5.4 to parameterize the WASMOD model for simulating the structural and functional variables of the seral stages. For each successional stage, there was a 20-year simulation of water and nutrient fluxes using meteorological data from 1970 to 1990. We tested the applicability of the model by comparing measured and simulated mean groundwater tables of the successional stages of the land use intensification sere. The measured and simulated results showed good agreement ($r^2 = 0.9$, $p > 0.001$; see also Müller *et al.* 2006).

5.3 Results

5.3.1 Changes of Ecosystem Features in the Land-Use Intensification Sere

The species richness of vascular plants in the land-use intensification sere (Fig. 5.4) was significantly higher in mesotrophic alder carrs (1A) than the eutrophic carrs (1B) (30 versus 24). Land-use intensification resulted in a drop in species numbers for the tall sedge reed communities (2A and 2B) with means of 15 and 8, respectively. After the initial drop there was an increase in species richness for successive stages, with a mean of 24 species in small sedge reeds (3) to 28 and 26 species in the wet meadows (4A and 4B). Thereafter, as land-use intensity increased, species richness decreased to a mean of 16 species in the wet pastures (5).

Simulated results for indicators of the nitrogen, carbon, and the efficiency measures are summarized in Table 5.5 and the following features are evident. Net nitrogen mineralization (NNM) is higher in alder carrs (1A and 1B; see Fig. 5.1) than in tall sedge reeds (2A and 2B), and remains low in unfertilized small sedge reeds (3) and mesotrophic wet meadows (4A). Higher mineralization rates up to 125 kg ha^{-1} a^{-1} are simulated for fertilized and more intensively drained eutrophic wet meadows and wet pastures (4B and 5). High N-leaching and denitrification rates are shown for these ecosystems. The other stages were low nitrogen sources. The highest net primary production (NPP) has been simulated for alder carrs (1A and 1B) and the lowest for small sedge reeds and mesotrophic wet meadows (3 and 4A). The simulated NPP of tall sedge reeds was similar to that of more intensively used wet grasslands. The simulations of the carbon budget suggest that alder carrs are carbon sinks (1A) or show a well-balanced carbon budget (1B), whereas eutrophic wet meadows and wet pastures can be characterized as high carbon sources. Alder carrs (1A and 1B) showed the highest values for the efficiency measures (NPP/MSR and NPP/Transpiration). Low values for NPP/MSR and NPP/T were detected for small sedge reeds (3) and the more intensively used wet meadows and pastures (4B and 5).

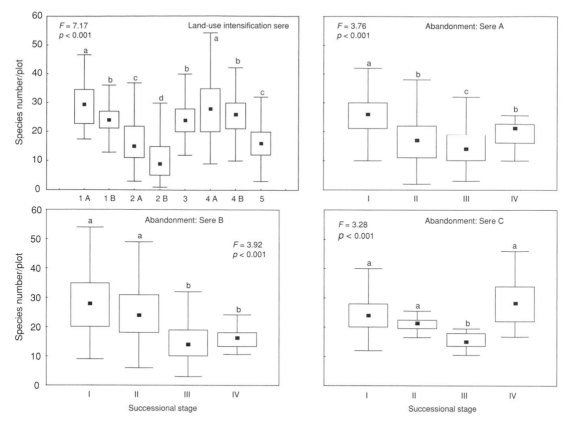

Figure 5.4 Change of species richness in the land-use intensification (stages 1A to 5, see Fig 5.1) and in stages I–IV for abandonment seres A, B and C. Plot size of alder carrs *c*. 100 m² and of types with herbaceous vegetation *c*. 16–25 m². Medians, 25 and 75% percentiles , and ranges without outliers (horizontal lines) are shown. Kruskal-Wallis-test (*F* and *p*) results are also shown. Different letters show significant differences ($p < 0.05$, Median test) across stages.

As a generalization across the retrogressive succession from alder carrs to wet pastures, C-balances progressively dropped; efficiency measures (NPP/MSR and NPP/Transpiration) also dropped but not as consistently as C-balances. Measures of N, hydrological budgets, and energy budgets started high (1A and 1B), then dropped (2A, 2B and 3), then rose again (4A, 4B and 5).

Table 5.5 Simulation results (WASMOD) for nitrogen and carbon budgets parameters, and efficiency measures in stages of the land use intensification sere. Ecosystem types are from Fig. 5.1.

Indicator:	**1A**	**1B**	**2A**	**2B**	**3**	**4A**	**4B**	**5**
				Ecosystem types				
NNM (kg N ha⁻¹ a⁻¹)	38	56	14	27	11	28	84	125
N-Leaching (kg N ha⁻¹ a⁻¹)	−15	−18	−9	−12	−6	−13	−46	−64
Denitr. (kg N ha⁻¹ a⁻¹)	17	29	3	7	3	12	52	100
N-balance (kg N ha⁻¹ a⁻¹)	−8	−23	3	−2	−4	−16	−24	−35
NPP (kg C ha⁻¹ a⁻¹)* 10³	4.8	6.0	2.9	4	1	2.1	3.0	3.5
C-balance (kg C ha⁻¹ a⁻¹)* 10³	+2	0	−0.6	+0.4	−1.3	−1.2	−5.6	−7.3
NPP/MSR	1.8	1.4	0.8	1.1	0.6	0.9	0.5	0.6
NPP/Transpiration	422	516	300	377	94	209	284	350

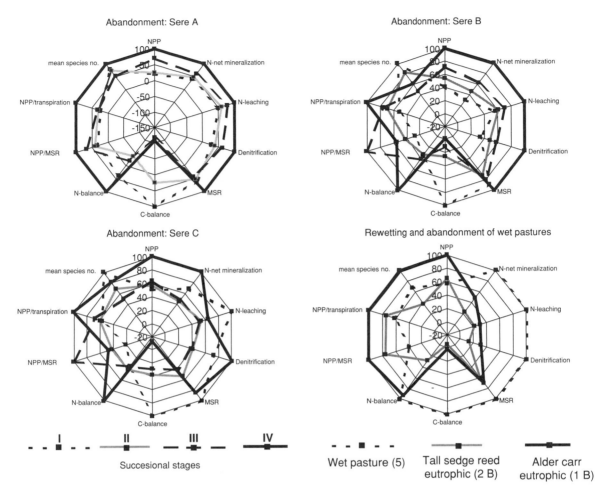

Figure 5.5 Amoeba diagrams indicate the changes of species richness and simulated indicators after abandonment of small sedge reeds (sere A), mesotrophic wet meadows (sere B), eutrophic wet meadows (sere C), and rewetting of wet pastures. In each sere, the highest values of the parameters were set as 100 (%). Negative values (C-, N-balances) characterize the systems as nutrient sinks.

5.3.2 Changes of Ecosystem Features in the Abandonment Seres

During succession, species richness decreased significantly in all abandonment seres from stages I to III (Fig. 5.4). Significant differences between stages I and II were only detected in successional sere A. In the seres A and C there was a significant increase of species richness from stage III to IV.

Amoeba diagrams (Fig. 5.5) illustrate the changes in all simulated parameters (mean values over the 20 years simulation period) for the abandonment succession for all seres and for rewetting abandoned wet pastures. In sere A all parameters except C-balance showed a progressive increase from stage 1 to stage 4. Such a clear-cut "progressive" trend was not found for all parameters in the other seres. In sere A (starting as the small sedge reed community 3) and B (starting as the mesotrophic wet meadow 4A) the NNM values increased from stage I to stage IV. In sere C (starting as the eutrophic wet meadow 4B), NNM values decreased from stage I to II and III, and then increased in stage IV. Thus,

Figure 5.6 Rewetted wet pastures, dominated by tall sedge reeds with *Carex acutiformis* and *Phalaris arundinacea* in Pohnsdorfer Stauung near Kiel, Germany.

in all seres, stage IV was the highest N source, and also had the highest denitrification values. N leaching in seres A and B increased from stage I to IV, but in sere C (wet eutrophic meadows 4B) N-leaching was highest in stage I. All seres showed a shift in C-balance from a carbon source (sedges/meadows/pastures) to a carbon sink (carrs). In all stages, the efficiency measures were highest in stage 4.

5.3.3 Change of Ecosystem Features After Rewetting

The development from intensively drained wet pastures (5) to eutrophic tall sedge reeds (2B) after rewetting (Fig. 5.6) resulted in a decrease of species richness (Fig. 5.5). However, species richness increased in the long run if the tall sedge reeds sere developed into eutrophic alder carrs (1B). Rewetting also led to a distinct decrease of NNM, N-leaching, denitrification rates, and microbial soil respiration (MSR). Furthermore, ecosystems changed from high to low nitrogen sources and from high to low carbon sources or carbon sinks. Values of efficiency measures (NPP/MSR, NPP/transpiration) were enhanced during succession after rewetting (Fig. 5.5).

5.3.4 Relationships Between Vegetation Structure, Plant Traits, and Species Groups

In stands of the abandonment fens the RI at 30 cm height decreased significantly with increasing standing crop ($r = 0.81$; $p < 0.001$). This was reflected in a positive correlation between the number of small-growing species and light availability ($r = 0.73$; $p < 0.001$) across the stages. Most of the *Scheuchzerio-Caricetea* species are small, whereas species of the *Phragmitetea*, that characterize tall sedge reeds and dominate in stages II and III of the successional seres, are larger.

Seeds collected from species of all successional stages were buried in 3 cm and 8 cm of litter in a greenhouse to test establishment characteristics. The results show positive relationships between seed mass and establishment; the greatest establishment was in 3 cm of litter (Fig. 5.7A and Fig. 5.7B). Seed weights of target species from the sociological groups characterizing small sedge reeds and wet meadows (*Scheuchzerio-Caricetea*, *Molinietalia*) varied over a wide range (Fig. 5.7C).

The longevity indices of Thompson *et al.* (1997) for seeds of these target species as well as species that are less specific to fen sites were between 0.3 and 0.5 (Fig. 5.7D). Although seeds from species of the *Molinietalia* had the lowest mean LI, the differences were not great and all species fell into the category "transient seed banks." Taking the results of the burial experiment into account, mean longevity indices of the same species groups were much higher. There were no significant differences between the groups (Fig. 5.7D).

5.3.5 Change of Species Richness by Mowing and Grazing

Results of field experiments in the wet meadow sites (4A and 4B) (Table 5.6) showed that after a period greater than 5 years species richness was increased by mowing once in summer and by mowing two times a year. Autumn mowing showed a smaller increase in species numbers, and mowing three to four times in summer showed a small decrease (Table 5.6). In the wet pastures (5), mowing in autumn and summer (one or two cuts) increased species numbers by a small amount. Species increases were relatively small and variation between plots was high, nevertheless the overall results show that mowing once in summer or mowing twice increased species numbers in the meadow and pasture plots (Fig. 5.8). In contrast, abandonment of meadows or pastures resulted in a significant decrease in species number.

To evaluate the success of management measures aimed at the increase of biodiversity, it is important to consider the dynamics of the target species (Fig. 5.9). Mesotropic and eutrophic wet meadows (4A and 4B) had increased cover of target species with mowing two times per year. The increases in cover values were evident after about 10 years. Numbers of target species remained relatively constant in the wet meadows during the investigation period. Reestablishment of mowing in late successional stages of eutrophic wet meadows and wet pastures

Table 5.6 Mean change of species number of wet meadows and wet pastures after establishment of different mowing regimes. In brackets: number of field experiments; x = not investigated. Ecosystems types are from Fig. 5.1.

Duration of experiment:	< 5 years		6–10 years		
Ecosystem type	4B	5	4A	4B	5
Abandonment	−2 (7)	−2 (7)	−5 (9)	−6 (8)	−2 (5)
Mowing (autumn, 1 cut)	x	0 (7)	−1 (13)	x	3 (5)
Mowing (summer, 1 cut)	5 (11)	2 (9)	2 (13)	5 (11)	4 (7)
Mowing (2 cuts)	4 (7)	1 (15)	3 (13)	x	3 (7)
Mowing (3–4 cuts)	2 (7)	x	-2 (13)	x	x

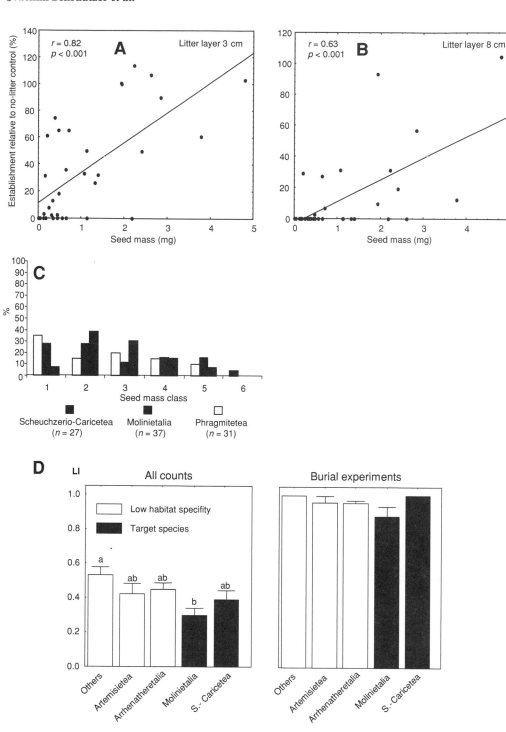

Figure 5.7 Relationships between seed mass (mg) and relative establishment of *Molinietalia* and *Scheuchzerio-Caricetea* species at a litter layer of 3 cm (A) and 8 cm (B). Establishment is given relative to a control treatment without a litter layer (Jensen and Gutekunst 2003). Percent of seeds in each of six seed mass classes (Grime *et al.* 1990) (C). Comparison of the Longevity index (LI) of different species groups (D) using all counts given in the data bank of Thompson *et al.* (1997) and results from burial experiments published in Jensen (2004).

Figure 5.8 Poorly drained mesotrophic wet meadow with *Dactylorhiza majalis*, regularly mowed once a year in late summer in Moenkeberger See near Kiel, Germany.

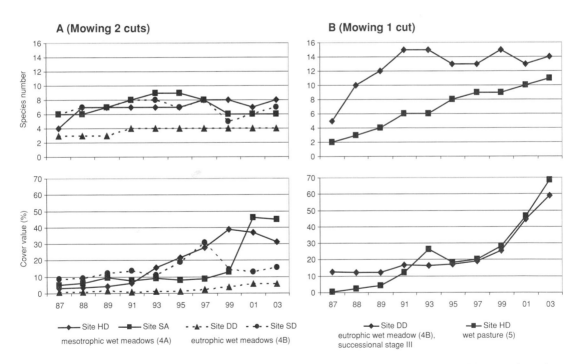

Figure 5.9 Development of target species (*Molinietalia, Scheuchzerio-Caricetea*) after establishment of mowing (A: 2 cuts, B: 1 cut). Site HD: Heubachwiesen; Site SA: Saerbeck; Site DD: Düsterdiek; Site SD: Strönfeld. Data: Biological Station Steinfurt.

Figure 5.10 Broadscale grazing in the Eider valley near Kiel, Germany.

led to a distinctive increase in numbers and cover values of target species already in the first 5 years of restoration (Fig. 5.9). However, we found no increase in rare species.

Our investigation in pastures of the Eider valley showed that increases in species richness by broadscale grazing (Fig. 5.10) depends on previous land-use and vegetation state (Table 5.7). In pasture I (dominantly wet meadow 4B, moderately grazed and not fertilized during the last several decades) species numbers started high and did not increase after 5 years. Pasture II (also 4B, previously heavily grazed but abandoned before the project started) had fewer species than pasture I at the start of the observations, but low intensity cattle grazing resulted in a significant increase in species richness. Species richness also increased in the previously intensively used pasture III (previously

Table 5.7 Development of species richness (plot size 625 m^2, $n = 7$) in different pastures of the Eider valley (Schleswig-Holstein) after implementation of broadscale grazing systems (1.5 cattle ha^{-1}, no fertilization). Pasture I: eutrophic wet meadows (4B), successional stages I and II, previous land-use: extensive grazing (1–2 cattle) without fertilization; Pasture II: eutrophic wet meadows (4B), successional stage III, previous land-use: abandonment; Pasture III: wet pastures (5), previous land-use: intensive grazing (3–4 cattle ha^{-1}) with fertilization. x = not investigated.

Year:	2000	2001	2002	2003	2004
Pasture I	58	57	55	58	53
Pasture II	31	30	34	38	40
Pasture III	x	x	31	38	41

intensively grazed, dominantly wet pastures) following the introduction of a moderate grazing regime.

5.4 Discussion

5.4.1 Effects of Land-Use Intensification on Ecosystem Features

Land-use intensity can be approximately equated with disturbance intensity (Grime 2001). However, here the mode of phytomass destruction indicating disturbance intensity varied (e.g., deforestation, mowing, or grazing) and other human impacts such as drainage and fertilization were superimposed. These complex variants need to be considered when evaluating the impact of disturbance intensity on ecosystem functioning. The intermediate disturbance hypothesis (IDH) is one of the most frequently suggested explanations for the coexistence of plant species in ecological communities (Wilson 1990). This hypothesis proposes that species richness is low at low and high disturbance, and is highest at medium disturbance. Species richness in the herbaceous stages of the land-use intensification sere from tall sedge reeds to wet pastures (Fig. 5.4; 2A through to 5) shows a hump-shape course conforming to the IDH (see also Gough *et al.* 1994, Schaffers 2002). The marked decline in species richness for the retrogressive succession from seminatural fen ecosystems (small sedge reeds and mesotrophic wet meadows) to wet pastures due to increased disturbance intensity has been noted in studies by Grootjans *et al.* (1986), Koerselman and Verhoeven (1995), and Wassen *et al.* (1996). However, if species-rich alder carrs are included, the IDH does not hold.

The set of ecosystem indicators for water and nitrogen budgets reveals no clear progressive trends for all indicators across the retrogressive stages, but C-balance and metabolic efficiency decreased. The patterns for indicators other than C-balance were mainly high to low to higher across the stages and did not correlate well with species richness. Among all stages, alder carrs had highest NPP and acted as carbon sinks or had a well-balanced carbon budget. Measurements concerning the carbon cycle of alder carrs are scarce. Kutsch *et al.* (2000) measured a mean NPP of 8002 kg C ha^{-1} a^{-1} and a C-balance of +3550 C kg ha^{-1} a^{-1} based upon measurements of all processes of the carbon cycle for this forest type. These values are about 30% higher than the mean values given by the simulation with WASMOD. The reason for this difference might be that the alder carr investigated by Kutsch *et al.* (2000) was at a relatively young stage of development.

The simulated NNM of alder carrs corresponded well with measured data. Döring-Mederake (1991) measured rates between 20 and 91 kg N ha^{-1} a^{-1}, which are in accordance with the range of our simulated values (25–80 kg N ha^{-1} a^{-1}). Due to higher simulated N-leaching and denitrification rates, alder carrs can be considered higher nitrogen sources than tall and small sedge reeds. Efficiency measures in alder carrs reached the highest values among all investigated systems.

Simulated NNM, N-leaching, and denitrification rates for tall and small sedge reeds (2A, 2B, and 3) were low and carbon as well as nitrogen balances were positive or weakly negative. Measured carbon balances of these systems were not available but measurements for the nitrogen budget (Koerselman and Verhoeven 1992) agreed with the simulated nitrogen balances.

Small sedge reeds (3) had low efficiency values, particularly the low ratio NPP/Transpiration. These low values suggest that species of small sedge reeds have to expend relatively high amounts of energy to take up nutrients at the nutrient-poor sites to produce biomass compared with species of the other ecosystems. These results conform to the classification of *Scheuchzerio-Caricetea* species as stress tolerators (Grime 2001). Higher values of the efficiency measures of tall sedge reeds (2A and 2B) that also grow at sites with relatively low nitrogen availabilities can be explained by the effective internal nutrient cycle of tall sedges. These species, classified mainly as competitive stress tolerators (Grime 2001), are able to translocate more than 50% of the nitrogen and phosphorus stored in the aboveground phytomass to their rhizomes at the end of the growing season (Denny 1987). These nutrients are remobilized in the following year and nutrients accumulate over several years, resulting in continuously increasing NPP rates. Unfertilized mesotrophic wet meadows (4A) can also be assessed as sustainable systems regarding the risk of nutrient losses: their simulated NNM, N-leaching, and denitrification rates were low and there were only low carbon and nitrogen sources.

The simulation results show that all parameters of the water and nutrient budgets develop negative values in fen ecosystems if drainage and fertilization are applied. Fertilized eutrophic wet meadows and especially wet pastures are ecosystems that bear the risk of higher nutrient losses to the atmosphere, to the groundwater, or to surface waters. In addition, the systems are high carbon and nitrogen sources. However, the simulated carbon balances (ca. -8000 kg C ha^{-1} a^{-1}) of the wet pasture might be overestimated as Kutsch *et al.* (2000) calculated only a carbon balance of -4000 kg C ha^{-1} a^{-1}. Hendriks (1993) investigated the nitrogen budget of a wet pasture in The Netherlands and detected a nitrogen balance of -60 kg N ha^{-1} a^{-1}; this is within the range of the simulated balances.

5.4.2 Effects of Abandonment on Ecosystem Features

Abandonment of seminatural fen ecosystems (small sedge reeds and wet meadows) causes a distinct decrease of species richness. This result is in accordance with many studies that were carried out in fen areas of northern and central Europe. The results reveal that abandonment mainly affects the decline of small species. Most *Scheuchzerio-Caricetea* species that dominate small sedge reeds and occur regularly in mesotrophic wet meadows belong to this species group. Almost half of these species are listed on endangered species lists of Germany. Two processes explain the extinction of small-growing species following abandonment. First, increasing standing crop during the succession reduces the light availability within the stands. The strong positive relationship between the number of small-growing species and light availability demonstrates that many of these species are strictly light demanding (Kotowski *et al.* 2001). The authors found that light intensity affected the growth of several *Scheuchzerio-Caricetea* species much more than the water level. In fens, different water levels can be also interpreted as differences in nitrogen availability (Okruszko 1993). The second process affecting the decline of species after abandonment is the development of a litter layer that hampers germination and establishment, especially of species with smaller seeds (Foster 1999, Jensen and Gutekunst 2003). However, we show here that this restriction refers to species of all sociological groups occurring on fen sites because these groups contain species with large as well as small seeds. The results of this study also reveal that in the long run, species richness could increase again if the systems develop to alder carrs.

Our simulation results have shown that the indicators of the carbon and nitrogen cycle and the nutrient balances show different patterns of change for the three successional seres selected. In the seres A and B, abandonment led to a continuous increase of NNM, N-leaching, and denitrification from stage I up to stage IV. In addition, microbial soil respiration increased. These differences were probably due to lower soil moisture caused by higher water used to support higher net primary production of the late successional stages. As a consequence, abandonment without rewetting could have negative effects on the functional properties of these systems. In the successional sere C, the decrease of fertilization reduces the risk of nutrient losses from the systems. Here the simulated N-leaching rate was much lower in successional stages II and III compared with stage I. On the other hand, it can be assumed, that the functional properties of the nitrogen budget would drastically get worse, if the systems change to drained alder carrs. NNM and N-leaching increased and N-balance became more negative. This effect can be explained by lower water tables in alder carrs compared with the intermediate successional stages of this sere.

The simulation results concerning the effects of abandonment on seminatural fen ecosystems are based only on changes of vegetation structure during succession. However, long-term physical changes to drained fen areas (such as a continuous closure of ditches by plants) can result in the rise of local water tables. These kinds of process were not considered in the simulations.

5.4.3 Effects of Rewetting on Ecosystem Structure and Function

Raising water levels up to the soil surface in drained wet pastures resulted in a decrease and later an increase in species richness. Roth *et al.* (1999) also observed the initial development from wet pastures to tall sedge reeds after rewetting. However, the success of rewetting strongly depends on the hydrological system and the quality of the available water (Grootjans *et al.* 2002). Rewetting intensively drained, eutrophic fen areas with precipitation water or stream water often results in the development of shallow lakes due to surface soil compaction. In this case, the succession will start with aquatic plants and will continue with the development of reeds beginning from the riparian zones of the lakes. In most areas, high nutrient concentrations in the stream water lead only to the establishment of eutrophic systems and common plant species. Asada *et al.* (2005) observed the expansion of the eutrophic reed species *Typha latifolia* in a flooded Canadian wetland consisting of bogs and surrounding drier peatlands after raising water levels up to 1.3 m. On the other hand, rewetting of wet pastures in discharge fen areas with deep, nutrient-poor groundwater enhances the potential for the establishment of mesotrophic fen species (Grootjans *et al.* 1996, Kieckbusch *et al.* 2006).

The simulation results indicate that raising water levels up to the soil surface in wet pastures efficiently reduces net nitrogen mineralization, nitrogen leaching, and microbial soil respiration if tall sedge reeds develop. Decreased nitrogen availability after rewetting due to lower mineralization rates are detected in the studies of Berendse *et al.* (1994), Updegraff *et al.* (1995), and van Duren *et al.* (1998; see Chapter 6), but other studies show no decrease of NNM after rewetting (van Dijk *et al.* 2004). The simulation results show that after rewetting wet pastures, a major change occurs in the carbon budget from a high carbon source to a carbon sink. How far the latter is realistic is difficult to assess, because peat-forming processes are dependent on the plant species that develop after rewetting (Roth et. al 1999) and the model WAS-MOD does not consider these aspects. According to Roth *et al.* (1999), only a few hygrophilous species such as *Carex elata*, *Carex acutiformis,* or *Phragmites australis* are peat-forming species. On the other hand, there are no data available that allow a reliable calculation of the carbon budget of tall sedge reeds because it seems to be nearly impossible to measure their belowground phytomass dynamics (Schrautzer 2004). Nevertheless, it can be assumed that the long-term development to eutrophic alder carrs will enhance the carbon storage capacity of rewetted fens (Kutsch *et al.* 2000). Apart from vegetation dynamics, the mode of rewetting also influences the carbon budget of the systems. Asada *et al.* (2005) demonstrated that dry peatlands might be changed to carbon sinks after flooding. The authors measured an accumulation of organic material of 10×10^3 kg ha^{-1} during a period of 9 years in areas with marsh vegetation.

The simulated denitrification rates of the rewetted ecosystems were relatively low. This was due to the fact that the simulation was carried out only at the site level and consequently, only internal processes were represented. However, rewetting of fens is often aimed at the reduction of nitrate concentrations of eutrophied surface water by denitrification (Leonardsson *et al.* 1994). The potential for denitrification of fen ecosystems is high and therefore the intensity of this process in rewetted fens depends mainly on the nitrate concentration of the surface water that is supplied (Davidsson *et al.* 2002). These aspects were not considered in our simulation.

5.4.4 Development of Species Richness After Mowing Without Fertilizing

Many characteristic species of seminatural fens such as small sedge reeds and mesotrophic wet meadows are missing in the agricultural landscape of northwestern Europe. Conservation of these species depends on a moderate disturbance regime to prevent successional changes caused by abandonment (Bakker and Berendse 1999, Jensen and Schrautzer 1999, Diemer *et al.* 2001). Moreover, to avoid the effects of eutrophication on species composition, external nutrient inputs, which in central Europe usually exceed the natural outputs of these low-productive systems, have to be controlled (Olde Venterink 2000). Mowing seems to be the most successful management measure to fulfill these demands (Grootjans and van Diggelen 1995; see Chapter 6).

The relationship between standing crop, light availability, and the occurrence of target species reveals that a specific ("target") biomass production of the stands is an important prerequisite to maintain or create suitable habitat conditions for many of these species (cf. Kotowski and van Diggelen 2004). Many light-demanding species of the *Scheuchzerio-Caricetea* occur mainly in stands with maximum standing crop values below 400 g m^{-2}, but the maintenance and reestablishment of most *Molinietalia* species do not require such low standing crop values. Investigations of Güsewell and Klötzli (1998) and Olde Venterink (2000) in other northwestern European fens as well as studies in

Canadian (Moore *et al.* 1989) and British fens (Wheeler and Shaw 1991) confirm this conclusion. Our results suggest that several species of the *Scheuchzerio-Caricetea* such as *Carex dioica* or *Eleocharis quinqueflora* even have optimal standing crop values lower than 200 g m^{-2} (cf. Schrautzer and Jensen 2006).

To moderate nutrient inputs by reducing standing crop in previously fertilized fen ecosystems it is important to know which nutrients actually limit aboveground phytomass production. A comparison of terrestrial wetlands (including fens) along a transect from western Europe to Siberia has shown that more endangered species persist under phosphorus-limited than under nitrogen-limited conditions (Wassen *et al.* 2005). The authors concluded that despite high N-deposition in western Europe, P-enrichment has been more accountable for the loss of wetland species than N-enrichment. As a consequence, one indispensable prerequisite for the maintenance of species-rich fens such as small sedge reeds or mesotrophic wet meadows is to prevent processes that enhance P-availability. Several studies have shown that in degraded fen ecosystems mostly potassium and sometimes phosphorus are the most important limiting nutrients (e.g., Schwartze 1992, Boeye *et al.* 1997). According to van Duren *et al.* (1998), yearly mowing led to a shortage of potassium and a reduction of aboveground phytomass in stands of drained fens, whereas the nitrogen availability remained high. A decrease in productivity of previously intensively used fens after long-term mowing has been observed in several other studies carried out in northwestern Europe (e.g., Bakker and Olff 1995).

Our literature survey about the success of haymaking experiments in degenerated fens confirmed that mowing once, or for best results twice a year, is a useful measure to maintain species-rich systems like mesotrophic wet meadows or to enhance the species richness of more degenerated systems. More frequent mowing will decrease species richness. The results of our own field experiments showed that percentage cover of target species increased after long-term mowing in all investigated systems. The potential for a reestablishment of these species seems to be highest in abandoned wet meadows (cf. Hald and Vinther 2000, Schwartze 2003). This might be explained by the conservation of long-term persistent seed banks due to the development of a decomposable litter layer. Previous investigations (Bekker *et al.* 1998b, Hölzel and Otte 2001) about the seed longevity of *Molinietalia* species concluded that many species only build up transient seed banks. As a consequence, the seed bank is considered to be unsuitable as a source for the reestablishment of wet meadow species in degenerated fens after restoration measures. Our seed burial experiments showed contrary results, with relatively long-term persistence in the peat. One important reason for these differences might be that the longevity of seeds was often underestimated using the determination method of Thompson *et al.* (1997). According to these authors, the seed longevity of species is classified as transient if the species occur in the current standing vegetation and could not be recorded in the soil samples. The latter might be the result of a low seed density in the soil. Bekker *et al.* (1997) mentioned that soil samples taken to analyze seed banks usually cover only 0.05% of the area that is used to record the current vegetation.

Although our field experiments showed that the abundance and number of target species increased during succession, other studies revealed no reestablishment of these species (e.g., Sach 1999, Grootjans *et al.* 2002). Consequently, the development potential of degenerated fens may be limited in many situations. Moreover, we found no field experiments in which hay-making of degenerated fens leads to the development of small sedge reeds with a full range of their characteristic (and mostly threatened) species. This phenomenon has complex reasons. First of all, the seed density in the soil generally decreases rapidly if the current vegetation does not continually add new seeds (Jensen 1998). In the case of *Scheuchzerio-Caricetea* species, short seed longevity cannot be used to explain the absence of these species as argued by other authors (Bekker *et al.* 1998b). The results of the burial experiments clearly reveal that most *Scheuchzerio-Caricetea* species build up a long-term persistent seed bank. Another reason for unsuccessful attempts to reestablish these species is that it is difficult to restore the hydrological system of fen areas (van Diggelen *et al.* 1994, Grootjans *et al.* 1996). Results of Runhaar *et al.* (1996) have shown that long-term hydrological changes at the landscape scale might be responsible for the decrease of target species in the long run as observed by Bakker (1989). Moreover, it should be taken into account that the changes of physical soil parameters at the sites caused by intensive drainage are almost irreversible (Zeitz 1992). The reversibility of compaction in drained peat soils after rewetting is low. Blankenburg *et al.* (2001) measured reswelling rates of 2–18 Vol. % in different peat soils.

5.4.5 Development of Species Richness After Grazing Without Fertilization

Extensive broadscale grazing has currently been introduced as a new strategy for the preservation of open landscapes in many parts of Europe (Finck *et al.* 2001). The main conservation objective of such projects is the development of a mosaic of different successional stages to offer suitable habitat conditions for many species and the development of high species richness. Moreover, broadscale grazing systems are considered as a cost-effective alternative to other management strategies such as mowing without fertilization (Härdtle *et al.* 2001). However, cattle grazing in species-rich wet grasslands have been rejected for many years as a useful measure to maintain species richness (Bakker and Grootjans 1991, Schrautzer and Wiebe 1993). One important argument used against extensive grazing in fen areas was that cattle prefer to graze dry places within the wetland, leading to undergrazing in wet areas, which promotes the development of tall-growing, species-poor vegetation.

The results presented here show that broadscale grazing maintains species richness in wet meadows and enhances the biodiversity of degraded fen ecosystems. However, the increase in species richness was due to increases in common grassland species. Grazing seems to be an effective management measure in species-rich, small sedge reeds as well. Based on the results of an indirect successional analysis, Stammel *et al.* (2003) detected that, despite a 15% lower mean species number in the grazed compared to the mown site, there were no differences in numbers of characteristic fen species.

Finally, it has to be taken into account that the limited reestablishment of target species in mowing and grazing experiments is probably also related to the absence of dispersal vectors in the cultural landscape of northwestern Europe and central Europe (Bakker *et al.* 1996). Dispersal of our target species is limited in this region by rare donor sites for seeds of these species (Bonn and Poschlod 1998). Furthermore, important dispersal agents such as regular flooding in river valleys are lacking due to the extensive construction of dykes. Based on the results of a greenhouse experiment, van den Broek *et al.* (2005) found that *Molinietalia* and *Scheuchzerio-Caricetea* species dispersal by hydrochory was restricted because of relative low buoyancy of their seeds. However, field investigations in the Eider valley and Estonian river valleys have shown that many *Molinietalia* and *Scheuchzerio-Caricetea* species are spread by hydrochory (unpublished data K. Voigt, A. Wanner, Ecology Centre of Kiel, Biological Institute University of Hamburg).

5.5 Conclusions and Recommendations for Restoration Management

In this chapter, we have described the species and abiotic changes in a retrogressive succession caused by increasing disturbance intensity from alder carrs to wet meadows, and a progressive succession for three retrogressive stages back to alder carrs. The results show no clear progressive changes in the functional indicators following land use intensification or abandonment. Successional changes and restoration actions are summarized in Fig. 5.11. Each of these systems fulfills important ecological functions in cultural landscapes developed over centuries. It seems unwise to favor any of these ecosystems over any other for nature protection. However, the development from low-productive mesotrophic wet meadows via eutrophic wet meadows to wet pastures supports the hypothesis of decreasing indicator values with increasing human impact and underlines the need to reduce land-use intensity in degraded ecosystems. During this intensification sequence (see also Müller *et al.* 2006) the indicators show the following differences during retrogression:

1. A reduction in biotic heterogeneity.
2. Exergy capture increases with rising productivity (which is the target of the dominant agricultural landscape management), while entropy production increases due to better conditions for microbial mineralization, i.e., after drainage.
3. Efficiency measures decrease with growing land-use intensity as do biotic and abiotic storage capacities, whereas nutrient loss is maximized by land-use intensification.

Abandonment of previous agricultural ecosystems often has been recommended as a nature protection "measure" in terms of reactivating self-organizational processes (Woodley *et al.* 1993, Jedicke 1995). If this were the case, then the indicators should generally show better values due to the reduced pressure, which potentially allows more pathways for self-organized dynamics. The results shown in Fig. 5.11 suggest that abandonment in fact improves the abiotic ecological functions of the ecosystems if successional stages develop that are characterized by clonal species such as tall sedges or large herbs. The initial decrease of NNM and nitrogen as well as carbon losses that might take place if eutrophic wet meadows are abandoned can be explained by the decrease of fertilization and rising water levels. However, the development of these

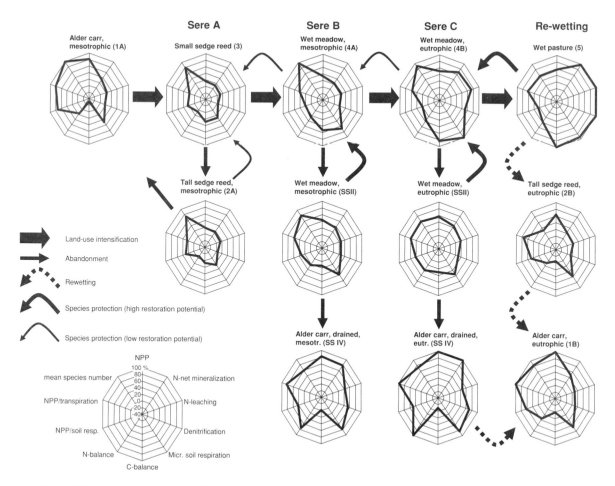

Figure 5.11 Amoeba diagrams to illustrate changes of the indicator values representing ecosystem functions during retrogressive succession following land-use intensification, and progressive succession following abandonment. Stages as in Fig. 5.1. Highest parameter values for all successional stages were set as 100%. Negative values (C-, N-balances) characterize the systems as nutrient sinks. Successional stages III of seres A, B, and C are not presented.

systems to alder carrs leads to higher nitrogen and carbon losses from the soil due to increased mineralization. The higher carbon storage in the woody phytomass of the ecosystems is restricted to the initial growth phase of these forests. It is difficult to predict when the development to alder carrs starts in abandoned wet meadows. Long-term observations have shown that late successional stages of abandoned wet meadows can remain stable for more than five decades due to missing disturbances. On the other hand, we detected a short-term development (5–10 years) of drained alder carrs mainly in previously grazed fens. These relationships should be taken into account if the implementation of large-scale grazing in fen areas is planned, because local undergrazing, which is a target of this nature conservation concept, can lead to a quick development of alder carrs (unpublished data, B. Holsten, Ecology Centre of Kiel). Furthermore, recommended management activities, such as planting alders to enhance carbon storage capacity of fens, has to be avoided if it is not possible to raise water levels in the degraded areas.

Implementation of appropriate restoration measures in highly degraded fen ecosystems such as wet pastures that cover more than 90% of the fen areas of most northern European countries is an important task within the scope of species and nature protection. Thus, raising the water levels in these systems to reduce their nutrient losses has to be a priority.

Nutrient budget simulations for an intensively utilized fen area (10% of the catchment area) and its surrounding mineral soils (90% of the catchment area) revealed that the fens contribute more than 90% to the nitrogen output of the catchment (Müller *et al.* 2006). Because of the high cost involved, the degree of rewetting usually depends on the socioeconomic conditions in the regions concerned. The most effective measure is to raise the water levels up to the soil surface that then leads to a decrease of NNM and microbial soil respiration. In the long run, it is possible to develop eutrophic wet alder carrs that have a higher species richness than the previously intensively used ecosystems (Fig. 5.11). Rewetting of fen areas with resulting initial development of shallow lakes is also a recommended measure if reduction of NNM and purification of polluted surface water is required. However, this measure bears the risk of increased phosphorus release (Grootjans *et al.* 2002, Kieckbusch *et al.* 2006).

If options to enhance water levels are restricted, measures that aim at the reestablishment of target species in strongly degraded fens should be introduced. Field experiments carried out in northwestern Europe have shown that grazing or mowing without fertilization usually enhances species richness and the number of target species (Fig. 5.11). Nevertheless, the success of these processes depends on the duration of these management measures. In our own field experiments, target species develop well in eutrophic wet meadows and wet pastures but not for up to 5–10 years. Moreover, the reestablishment of target species is restricted to *Molinietalia* species and common *Scheuchzerio-Caricetea* species. In contrast, the potential for reestablishment of rare species is low due usually to missing dispersal vectors, lack of suitable donor sites, and far-reaching, irreversible soil compaction. Removal of the upper, degraded soil horizon (sod cutting) is often recommended as a small-scale restoration measure in highly degraded fens. This measure will enhance habitat conditions for target species, but is too costly for broadscale use. Consequently, species protection in fens of northwestern Europe should mainly focus on the maintenance of species-rich small sedge reeds and wet meadows.

Finally, our results show both irreversible changes on the one hand and hysteresis effects on the other. Abandonment and self-organizing processes will not always develop a desired successional stage and technical measures such as rewetting in abandoned wet pastures are needed.

Acknowledgements: The authors thank Rudy van Diggelen, Werner Härdtle, David Mackenzie, Roger del Moral, Gert Rosenthal, and especially Joe Walker for careful reviews of the manuscript.

References

Asada, T., Warner, B. G., and Schiff, S. L. 2005. Effects of shallow flooding on vegetation and carbon pools in boreal wetlands. *Applied Vegetation Science* 8:199–208.

Bakker, J. P. 1989. *Nature Management by Grazing and Cutting*. Dordrecht: Kluwer.

Bakker, J. P., and Berendse, F. 1999. Constraints in the restoration of ecological diversity in grassland and heathland communities. *Trends in Ecology and Evolution* 14:63–68.

Bakker, J. P., and Grootjans, A. P. 1991. Potential for vegetation regeneration in the middle course of the Drentsche A Brook valley (The Netherlands). *Verhandlungen der Gesellschaft für Ökologie* 20:249–263.

Bakker, J. P., and Olff, H. 1995. Nutrient dynamics during restoration of fen meadows by hay making without fertilizer application. In: *Restoration of Temperate Wetlands*. B. D. Wheeler, S. C., Shaw, W. J., Foijt, and A. Robertson (eds.). Chichester: Wiley and Sons, pp. 143–166.

Bakker, J. P., Poschlod, P., Strijkstra, R. J., Bekker, R. M., and Thompson, K. 1996. Seed bank dynamics and seed dispersal: Important topics in restoration ecology. *Acta Botanica Neerlandica* 47:15–26.

Bekker, R. M., Verweij, G. L., Smith, R. E. N., Reine, R., Bakker, J. P., and Sneider, S. 1997. Soil seed banks in European grasslands: Does land use affect regeneration perspectives? *Journal of Applied Ecology* 34:1293–1310.

Bekker, R. M., Bakker, J. P., Grandin, U., Kalamees, R., Milberg, P., Poschlod, P., Thompson, K., and Willems, J. H. 1998a. Seed size, shape and vertical distribution in the soil: Indicators of seed longevity. *Functional Ecology* 12:834–842.

Bekker, R. M., Schaminee, J. H. J., Bakker, J. P., and Thompson, K. 1998b. Seed bank characteristics of Dutch plant communities. *Acta Botanica Neerlandica* 47:15–26.

Berendse, F., Oomes, M. J. M., Altena, H. J., and de Visser, W. 1994. A comparative study of nitrogen flows in two similar meadows affected by different groundwater levels. *Journal of Applied Ecology* 31:40–48.

Blankenburg, J., Hennings, H. H., and Schmidt, W. 2001. Bodenphysikalische Eigenschaften von Torfen und Wiedervernässung. In: *Ökosystemmanagement Für Niedermoore. Strategien und Verfahren zur Renaturierung*. R. Kratz and J. Pfadenhauer (eds.). Stuttgart, Germany: Eugen Ulmer, pp. 81–91.

Boeye, D., Verhagen, B., van Haesebroeck, V., and Verheyen, R. F. 1997. Nutrient limitation in species-rich lowland fens. *Journal of Vegetation Science* 8:415–425.

Bonn, S., and Poschlod, P. 1998. *Ausbreitungsbiologie der Pflanzen Mitteleuropas*. Wiesbaden, Germany: Quelle and Meyer.

Bragg, O., and Lindsay, R. (eds.). 2003. *Strategy and Action Plan for Mire and Peatland Conservation in Central Europe*. Wageningen, The Netherlands: Wetlands International.

Davidsson, T., Trepel, M., and Schrautzer, J. 2002. Denitrification in drained and rewetted minerotrophic peat soils in Northern Germany (Pohnsdorfer Stauung). *Journal of Plant Nutrition and Soil Science* 165:199–204.

Denny, P. 1987. Mineral cycling by wetland plants. A review. *Archiv Hydrobiologische Beihefte* 27:1–25.

Diemer, M., Oetiker, K., and Billeter, R. 2001. Abandonment alters community composition and canopy structure of Swiss calcareous fens. *Applied Vegetation Science* 4:237–246.

Döring-Mederake, U. 1991. Feuchtwälder im nordwestdeutschen Tiefland. Gliederung—Ökologie—Schutz. Göttingen: Scripta Geobotanica 19.

EPA (Environmental Protection Agency of the United States) 2006. Description of wetland types. [http://epa.gov/owow/wetlands/types/fen.html].

Finck, P., Riecken, U., and Schröder, E. 2001. Pasture landscapes and nature conservation. New strategies for preservation of open landscapes in Europe. In: *Pasture Landscapes and Nature Conservation*. B. Redecker, P., Finck, U., Riecken, and W. Härdtle (eds.). Berlin: Springer, pp. 1–14.

Foster, B. L. 1999. Establishment, competition and the distribution of native grasses among Michigan old-fields. *Journal of Ecology* 87:476–489.

Fraser, L. H. and Keddy, P. A. 2005. *The World's Largest Wetlands—Ecology and Conservation*. Cambridge: Cambridge University Press.

GEC (Global Environment Centre) 2003. ASEAN Peatland Management Initiative. Sustainable Management of Peatlands: Wise Use, Prevention of Fires and Rehabilitation of Peatland. [www.peat-portal.net]

Gough, L., Grace. J. B., and Taylor, K. L. 1994. The relationships between species richness and community biomass: the importance of environmental variables. *Oikos* 70:271–279.

Grime, J. P., Hodgson, J. G., and Hunt, R. 1990. *The Abridged Comparative Plant Ecology*. London: Chapman and Hall.

Grime, J. P. 2001. *Plant Strategies, Vegetation Processes, and Ecosystem Properties*. Chichester: Wiley.

Grootjans, A. P., and van Diggelen, R. 1995. Assessing the restoration prospects of degraded fens. In: *Restoration of Temperate Wetlands*. B. D. Wheeler, S. C. Shaw, W. J. Foijtt, and A. Robertson (eds.). Chichester: Wiley, pp. 73–90.

Grootjans, A. P., Schipper, P. C., and van der Windt, H. J. 1986. Influence of drainage on N-mineralisation and vegetation response in wet meadows. II: Cirsio-Molinietum stands. *Oecologia Plantarum* 7:3–14.

Grootjans, A.P., van Wirdum, G., Kemmers, R., and van Diggelen, R. 1996. Ecohydrology in the Netherlands: Principles of an application-driven interdiscipline. *Acta Botanica Neerlandica* 45:491–516.

Grootjans, A. P., Bakker, J. P., Jansen, A. J. M., and Kemmers, R. H. 2002. Restoration of brook valley meadows in The Netherlands. *Hydrobiologia* 478:149–170.

Güsewell, S., and Klötzli, F. 1998. Abundance of common reed (*Phragmites australis*), site conditions and conservation value of fen meadows in Switzerland. *Acta Botanica Neerlandica* 47:113–129.

Hald, A. B., and Vinther, E. 2000. Restoration of a species-rich fen-meadow after abandonment: Response of 64 plant species to management. *Applied Vegetation Science* 3:15–24.

Härdtle, W., Mierwald, U., Behrends, T., Eischeid, I., Garniel, U., Grell, H., Haese, D., Schneider-Fenske, A., and Voigt, N. 2001. Pasture landscapes in Germany—progress towards sustainable use of agricultural land. In: *Pasture Landscapes and Nature Conservation*. B. Redecker, P. Finck, U. Riecken, and W. Härdtle (eds.). Berlin: Springer, pp. 147–160.

Hendriks, R. F. A. 1993. Nutrientenbelasting van oppervlaktewater in veenweidegebieden. Wageningen, DLO-Staring Centrum, Rapport 251.

Hölzel, N., and Otte, A. 2001. The impact of flooding regime on the soil seed bank of flood-meadows. *Journal of Vegetation Science* 12:209–218.

Jedicke, E. 1995. Ressourcenschutz und Prozessschutz. Diskussion notwendiger Ansätze zu einem ganzheitlichen Naturschutz. *Naturschutz und Landschaftsplanung* 27:125–133.

Jensen, K. 1998. Species composition of soil seed bank and seed rain of abandoned wet meadows and their relation to aboveground vegetation. *Flora* 139:345–359.

Jensen, K. 2004. Langlebigkeit der Diasporenbanken von Arten der Niedermoorflora Nordwest-Deutschlands: Überblick und Methodenvergleich. *Berichte der Reinhold-Tüxen-Gesellschaft* 16:17–28.

Jensen, K., and Schrautzer, J. 1999. Consequences of abandonment for a regional fen flora and mechanism of succession change. *Applied Vegetation Science* 2:79–88.

Jensen, K., and Gutekunst, K. 2003. Effects of litter on establishment of grassland plant species: The role of seed size and successional status. *Basic and Applied Ecology* 4:579–587.

Joergensen, S. E. 2001. A general outline of thermodynamic approaches to ecosystem theory. In: *Handbook of Ecosystem Theories and Management*. S. E. Joergensen and F. Müller (eds.). New York: CRC Publishers, pp. 113–134.

Joosten, H., and Couwenberg, J. 2001. Bilanzen zum Moorverlust – das Beispiel Europa. In: *Landschaftsökologische Moorkunde*. M. Succow and H. Joosten (eds.). Stuttgart, Germany: Schweizerbart'sche Verlagsbuchhandlung, pp. 406–409.

Kieckbusch, J., Schrautzer, J., and Trepel, M. 2006. Spatial heterogeneity of water pathways in degenerated riverine peatlands. *Basic and Applied Ecology* 7:388–397.

Koerselman, W., and Verhoeven, J. T. A. 1992. Nutrient dynamics in mires of various status: Nutrient inputs and outputs and the internal nutrient cycle. In: *Fens and Bogs in The Netherlands: Vegetation, History, Nutrient Dynamics and Conservation*. J. T. A. Verhoeven (ed.). Dordrecht: Kluwer, pp. 397–432.

Koerselman, W., and Verhoeven, J. T. A. 1995. Eutrophication of fen ecosystems: External and internal nutrient sources and restoration strategies. In: *Restoration of Temperate Wetlands*. B. D. Wheeler, S. C. Shaw, J. W. Foijtt, and A. Robertson (eds.). Chichester: Wiley, pp. 91–112.

Kotowski, W., van Andel, J., van Diggelen, R., and Hogendorf, J. 2001. Responses of fen plant species to groundwater level and light intensity. *Plant Ecology* 155:147–156.

Kotowski, W., and van Diggelen, R. 2004. Light as an environmental filter in fen vegetation. *Journal of Vegetation Science* 15:583–594.

Kutsch, W., Dilly, O., Steinborn, W., and Müller, F. 1998. Quantifying ecosystem maturity—a case study. In: *Eco Targets, Goal Functions and Orientors*. F. Müller and M. Leupelt (eds.). New York: Springer, pp. 209–231.

Kutsch, W., Eschenbach, C., Dilly, O., Middelhoff, U., Steinborn, W., Vanselow, R., Weisheit, K., Wötzel, J., and Kappen, L. 2000. The carbon cycle of contrasting landscape elements of the Bornhöved Lake District. *Ecological Studies* 147:75–95.

Kutsch, W., Steinborn, W., Herbst, M., Baumann, R., Barkmann, J., and Kappen, L. 2001. Environmental indication: A field test of an ecosystem approach to quantify biological self-organization. *Ecosystems* 4:49–66.

Leonardsson, L., Bengtson, L., Davidsson, T., Persson, T., and Emanuelsson, U. 1994. Nitrogen retention in artificially flooded meadows. *Ambio* 23:332–334.

Mitsch, W.J., and Gosselink, J. G. 2000. *Wetlands*. 3rd edn. New York: Wiley.

Moore, D. R. J., Keddy, P. A., Gaudet, C. L., and Wisheu, I. C. 1989. Conservation of wetlands: Do infertile wetlands deserve a higher priority? *Biological Conservation* 47:203–217.

Müller, F. 1998. Gradients, potentials and flows in ecological systems. *Ecological Modelling* 108:3–21.

Müller, F., and Fath, B. 1998. The physical basis of ecological goal functions—An Integrative Discussion. In: *Eco Targets, Goal Functions, and Orientors*. F. Müller and M. Leupelt (eds.). New York: Springer, pp. 269–285.

Müller, F., and Joergensen, S. E. 2000. Ecological orientors: A path to environmental applications of ecosystem theories. In: *Handbook of Ecosystem Theories and Management*. S. E. Joergensen and F. Müller (eds.). New York: CRC Publishers, pp. 561–576.

Müller, F., Schrautzer, J., Reiche, E. W., and Rinker, A. 2006: Ecosystem based indicators in retrogressive successions of an agricultural landscape. In: *Ecological Indicators* 6:63–82.

Odum, H. T., Brown, M. T., and Ulgiati, S. 2000. Ecosystems as energetic systems. In: *Handbook of Ecosystem Theories and Management*. S. E. Joergensen and F. Müller (eds.). Boca Raton, Florida: Lewis Publishers.

Okruszko, H. 1993. Transformation of fen-peat soil under the impact of draining. *Zeszyty Problemowe Postepow Nauk Rolniczych* 406:3–73.

Olde Venterink, H. 2000. Nitrogen, phosphorus and potassium flows controlling plant productivity and species richness; eutrophication and nature management in fens and meadows. PhD Thesis, Utrecht University.

Pfadenhauer, J., and Grootjans, A. P. 1999. Wetland restoration in Central Europe. Aims and methods. *Applied Vegetation Science* 2:95–106.

Ramsar Convention 2002. Resolution VIII.17 on Global Action on Peatlands. [www.ramsar.org/key_res_viii_17_e.html]

Reiche, E. W. 1994. Modelling water and nitrogen dynamics on catchment scale. *Ecological Modelling* 75/76:372–384.

Rosenthal, G., Hildebrandt, J., Zöckler, C., Hengstenberg, M., Mossakowski, D., Lakomy, W., and Burfeindt, I. 1998. *Feuchtgrünland in Nordeutschland. Ökologie, Zustand, Schutzkonzepte*. Münster, Germany: Bundesamt für Naturschutz Schriftenvertrieb.

Roth, S., Seeger, T., Poschlod, P., Pfadenhauer, J., and Succow, M. 1999. Establishment of halophytes in the course of fen restoration. *Applied Vegetation Science* 2:131–136.

Runhaar, J., van Gool, C. R., and Groen, C. L. G. 1996. Impact of hydrological changes on nature conservation areas in The Netherlands. *Biological Conservation* 76:269–347.

Sach, W. 1999. Vegetation und Nährstoffdynamik unterschiedlich genutzten Grünlandes in Schleswig-Holstein. Dissertationes Botanicae 308, Stuttgart.

Schaffers, A. P. 2002. Soil, biomass and management of semi-natural vegetation. Part II. Factors controlling species diversity. *Plant Ecology* 158:247–268.

Schieferstein, B. 1997. Ökologische und molekularbiologische Untersuchungen am Schilf (*Phragmites australis* (Cav.) Trin. ex Staud.) im Bereich der Bornhöveder Seenkette. EcoSys 22, Kiel.

Schneider, E. D. and Kay, J. J. 1994. Life as a manifestation of the second law of thermodynamics. *Mathematical and Computer Modelling* 19:25–48.

Schrautzer, J. 1988. Pflanzensoziologische und standörtliche Charakteristik von Seggenriedern und Feuchtwiesen in Schleswig-Holstein. Mitteilungen der AG Geobotanik in Schleswig-Holstein und Hamburg 38, Kiel.

Schrautzer, J. 2004. Niedermoore Schleswig-Holsteins: Charakterisierung und Beurteilung ihrer Funktion im Landschaftshaushalt. Mitteilungen der AG Geobotanik in Schleswig-Holstein und Hamburg 63, Kiel.

Schrautzer, J., and Wiebe, C. 1993. Geobotanische Charakterisierung und Entwicklung des Grünlandes in Schleswig-Holstein. *Phytocoenologica* 22:105–144.

Schrautzer, J., Asshoff, M., and Müller, F. 1996. Restoration strategies for wet grasslands in Northern Germany. *Ecological Engineering* 7:255–278.

Schrautzer, J., and Jensen, K. 2006. Relationship between light availability and species richness during fen grassland succession. *Nordic Journal of Botany* (in press).

Schwartze, P. 1992. Nordwestdeutsche Feuchtgrünlandgesellschaften unter kontrollierten Nutzungsbedingungen. Dissertationes Botanicae 183, Stuttgart.

Schwartze, P. 2003. Einfluss von Brache und Nutzung auf Feuchtgrünlandvegetation im Münsterland. Kieler Notizen zur Pflanzenkunde. *Schleswig-Holstein Hamburg* 31:185–196.

Stammel, B., Kiehl, K., and Pfadenhauer, J. 2003. Alternative management on fens: Response of vegetation to grazing and mowing. *Applied Vegetation Science* 6:245–254.

Svirezhev, Y. M., and Steinborn, W. 2001. Exergy of solar radiation: Thermodynamic approach. *Ecological Modelling* 145:101–110.

Thompson, K., Bakker, J. P., and Bekker, R. M. 1997. *The Soil Seed Banks of North West Europe: Methodology, Density and Longevity*. Cambridge: University of Cambridge Press.

Trepel, M. 2000. Quantifizierung der Stickstoffdynamik von Ökosystemen auf Niedermoorböden mit dem Modellsystem WASMOD. EcoSys 29, Kiel.

Updegraff, K., Bridgham, S. D., Pastor, J., and Johnston, C. A. 1995. Environmental and substrate controls over carbon and nitrogen mineralization in northern wetlands. *Ecological Applications* 5:151–163.

van den Broek, T., van Diggelen, R., and Bobbink, R. 2005. Variation in seed buoyancy of species in wetland ecosystems with different flooding dynamics. *Journal of Vegetation Science* 16:579–586.

van Diggelen, R., Grootjans, A. P., and Burkunk, R. 1994. Assessing restoration perspectives of disturbed brook valleys: The Gorecht area, The Netherlands. *Restoration Ecology* 2:87–96.

van Dijk, J., Stroetenga, M., Bos, L., van Bodgom, P. M., Verhoef, H. A., and Aerts, R. 2004. Restoring natural seepage conditions on former agricultural grassland does not lead to reduction of organic matter decomposition and soil nutrient dynamics. *Biogeochemistry* 71:317–337.

van Duren, I. C., Boeye, D., and Grootjans, A. P. 1998. Nutrient limitations in an extant and drained poor fen: Implications for restoration. *Plant Ecology* 133:91–100.

van Duren, I. C., Strijkstra, R. J., Grootjans, A. P., ter Heerdt, G. J. N., and Pegtel, D. M. 1998. A multidisciplinary evaluation of restoration measures in a degraded Cirsio-Molinietum fen meadow. *Applied Vegetation Science* 1:115–130.

Wassen, M. J., van Diggelen, R., Verhoeven, J. T. A., and Wolejko, L. 1996. A comparison of fens in natural and artificial landscapes. *Vegetatio* 126:5–26.

Wassen, M. J., Olde Venterink, H., Lapshina, E. D., and Tanneberger, F. 2005. Endangered plants persist under phosphorus limitation. *Nature* 437:547–550.

Wetlands International Russia. 2003. *Action Plan for Peatland Conservation and Wise Use in Russia.* Moscow, Russia: Wetlands International Russia Programme.

Wheeler, B. D., and Shaw, S. C. 1991. Above-ground crop mass and species richness of the principle types of herbaceous rich-fen vegetation of lowland England and Wales. *Journal of Ecology* 79:285–301.

Wilson, J. B. 1990. Mechanisms of species coexistence: Twelve explanations for Hutchinson's "paradox of the plankton": Evidence from New Zealand plant communities. *New Zealand Journal of Ecology* 18:176–181.

Woodley, S., Kay, J., and Francis, G. 1993. *Ecological Integrity and the Management of Ecosystems.* Ottawa, Canada: St. Lucie Press.

Zeitz, J. 1992. Bodenphysikalische Eigenschaften von Substrat-Horizont-Gruppen in landwirtschaftlich genutzten Niedermooren. *Zeitschrift für Kulturtechnik und Landentwicklung* 33:301–307.

6

Manipulation of Succession

Karel Prach, Rob Marrs, Petr Pyšek, and Rudy van Diggelen

Key Points

1. Ecological restoration can be achieved using either unassisted succession, a manipulation of spontaneous succession, or technical restoration. We describe each of these approaches and suggest under what circumstances each of them can be used.
2. There are two principal directions by which succession can be manipulated to attain a target, either to accelerate it or to reverse it if it has proceeded beyond the target. Manipulation of both the physical environment and the biota are considered.
3. Examples are given from mining sites, abandoned fields, secondary grasslands, heathlands, and wetlands in Europe.

6.1 Introduction

Succession comprises many ecological processes that underpin all ecological restoration and ecological restoration is a manipulation of these processes to achieve its goals. This means it is essential to understand how succession operates, and when and how to manipulate it. The main goals of manipulating succession are to (i) increase the natural value of degraded ecosystems; this goal is often restricted to an effort to increase species diversity (Perrow and Davy 2002) but it may not be desirable if, for example, alien species are a component of the increased diversity; (ii) increase ecosystem productivity, which is important in those parts of the world where any increase in productivity is desirable from a socioeconomic perspective (Wali 1992, Whisenant 1999); (iii) increase ecosystem services, for example, to protect against soil erosion, erect buffer zones against pollution, or to improve the aesthetic quality of a site (van Andel and Aronson 2006). A fundamental starting point for any restoration scheme is to define both the starting conditions and the target ecosystem or endpoint. In this chapter, we will restrict our attention to targets with a high natural value, where successional processes and their manipulation are important, and ignore the productivity and service targets. We attempt to address the question: Under what circumstances is it possible to rely on unassisted spontaneous succession? Alternatively, when is manipulation of spontaneous succession or technical reclamation needed to reach the targets?

6.2 Manipulation of Succession: A Framework

6.2.1 Spatial Scales

Two types of information can be obtained from restoration schemes where succession has been manipulated: The first is from controlled experiments that are typically small-scale (10^{-1}–10^2 m^2) and last only a few years. Resulting data can be tested rigorously, and have a potential to contribute to ecological theory. However, their extrapolation to the scales at which most practical restoration projects are implemented (10^3–10^5 m^2) is limited, and must be made with caution (see Chapter 1). Second, there are general observations that are obtained from practical restoration work. Such information is typically available at larger scales but must also be used with caution because it cannot be tested rigorously. Nevertheless, this observational information can be used to generate hypotheses for subsequent testing and verification as well as in the interpretation of the experimental data. Each restoration project ought to use both types of knowledge.

Inevitably a great influence on any restoration scheme is the landscape context, and how the treated area relates to its surroundings. An important constraint on succession is the available species pool, which is determined by a combination of factors, such as macro- and microclimate, areas of intact vegetation, land-use history (Zobel *et al.* 1998), and by their spatial patterns. Propagule sources in the close vicinity are important for the establishment of late-successional species that usually have poor dispersal abilities (Poulin *et al.* 1999, Novák and Prach 2003; see Chapter 2). Generalists, often present among early-successional species, disperse more readily, and can colonize from a distance (Grime 1979). Their participation in succession is often determined by their abundance in the surrounding landscape and a mass effect is important (Settele *et al.* 1996). On the other hand, seral stages on restored sites may serve as propagule sources of generalists, including invasive aliens and weeds, for the surrounding landscape (Rejmánek 1989). Various human activities in the surrounding landscape influence the regional species pool and thus influence vegetation succession in a restored site. Any intentional manipulation of propagule sources in the wider landscape is, however, difficult. This is particularly true for generalists, including aliens, with widely dispersed propagules (see Chapter 5).

6.2.2 Temporal Scales

6.2.2.1 *Rate of Succession*

In human-altered landscapes, recovery to late-successional communities with a slow species turnover can take many decades, yet these communities are often used as targets in restoration programs and succession is manipulated to accelerate succession toward them. Under extreme site conditions, such as china clay wastes in south-west England, unassisted succession can take a very long time (>100 years) and patches devoid of vegetation can persist (Marrs and Bradshaw 1993). At this site, technical manipulations can reduce the process of succession to a sustainable ecosystem in as less as 7 years. However, in many cases unassisted succession can produce late-successional stages in a reasonable time. For example, in various human-made sites in central Europe, such stages develop quite spontaneously after 20–30 years since abandonment, which is an acceptable time for restoration purposes (Prach and Pyšek 2001). On the other hand, restoration as a reverse process from the late-successional communities toward younger ones may also be considered, thus going against the natural direction and rate of succession.

6.2.2.2 Role of Timing

The impact of timing of manipulation depends to a large extent on the treatment being used and the system being manipulated. For example, when managing succession to establish/maintain moorlands with a dominant *Calluna vulgaris,* a burning/cutting frequency of 6–14 years would be appropriate (Gimingham 1992), but for a species-rich grassland flora annual mowing is needed (WallisDeVries *et al.* 1998). In a 10-year experiment to control a late-successional species, *Pteridium aquilinum*, in a heathland, herbicide treatment applied in the first 2 years gave a rapid reduction in *P. aquilinum*, but the reduction in plant density was temporary. Annual cutting treatments were needed to maintain the heathland (Tong *et al.*, 2006).

Manipulation applied at different times of the growing season may have very different and often opposing effects. For example, dominant grasses can be maintained by cutting late in the season, when reserves have been translocated to underground organs, or suppressed if the cutting is applied early in the season (Klimeš and Klimešová 2002). Similarly, spring grazing of calcareous grassland in England led to a higher species diversity and vegetation containing more target species than autumn grazing (Gibson *et al.* 1987). Eradication of invasive species is usually effective only if conducted at an appropriate phenological phase, usually at the time of their intensive growth and, of course, before the invasive species set seed (Pyšek *et al.* 1995).

Seeding of target species must be carried out in an appropriate "colonization window," the duration of which depends on the species and sere (Johnstone 1986). It is logical within a restoration scheme to use these colonization windows to maximum benefit. The opportunity for intervention appears to occur at the point when dominant species or life forms change due to spontaneous processes (Prach 2003). These colonization windows can be influenced dramatically by extreme weather events (Marrs and Le Duc 2000, Bartha *et al.* 2003).

In some restoration projects it is reasonable to distinguish between short-term and long-term restoration goals, but these may conflict. For example, planting alien ground-cover species to minimize short-term erosion slowed down long-term restoration of target vegetation in coal mine sites (Ninot *et al.* 2001, Holl 2002).

6.2.3 Position of a Restored Site on Environmental Gradients

The effort to find environmental variables best correlated with successional pattern is as old as studies on succession. However, it is difficult to find clear and generally valid correlations, despite the commonly accepted role of climate (Morecroft *et al.* 2004), site moisture, and nutrients (Tilman 1988). Soil pH is often a useful predictor of vegetation succession in terms of species composition (Christensen and Peet 1984), and it was the only soil characteristic significantly affecting successional pattern in a comparative study of 15 unassisted successions in various human-made habitats in central Europe (Prach *et al.* unpubl.). These studies suggest that manipulation of soil pH might be a useful tool for accelerating and directing succession toward specific targets.

In many systems, low nutrient availability assists restoration because it restricts the growth of competitive, nontarget species. This generalization appears valid for nutrient-rich landscapes such as in temperate Europe, but not in extreme, marginal areas (Whisenant 1999). Low-fertility substrates often provide good establishment opportunities for those species that are weak competitors on fertile sites. Thus, these infertile sites can serve as refugia for rare and endangered species retreating from nutrient-enriched landscapes (Benkewitz *et al.* 2002, Pywell *et al.* 2003). Nutrient levels can be easily increased, but not so easily reduced by manipulation (Perrow and Davy 2002).

6.3 Methodological Approaches

6.3.1 Moving Succession Toward a Target

Three different strategies can be envisaged for creating new ecosystems during ecological restoration, representing a gradient of management and intervention intensity to manipulate succession:

(i) The simplest approach is to leave the site without intervention (unassisted succession; Parker 1997, Prach *et al.* 2001, Walker and del Moral 2003); successional development then proceeds at its own pace, but will be substantially affected by the local species pool. This strategy is sometimes slow, taking decades or even centuries, and it is often difficult to ensure that the final target ecosystem is met. However, costs to implement this strategy are low as long as no goods or services are affected.

(ii) At the other end of the spectrum is a technical solution to reach a target. Here, many abiotic variables can be altered, and biota and biotic processes can be more or less controlled by introducing desirable species. While there is still a role for colonizing biota, it is likely that locally derived biota will have a reduced importance as establishment into developed or developing vegetation may be more difficult due to additional competition from sown or planted species.

(iii) Between these extremes there is an approach where spontaneous succession is assisted by limited physical or biotic manipulations. Physical manipulation of succession may rely on improving the site and then allowing spontaneous succession, i.e., relying on colonization processes to create the community species pool. With biotic manipulation, adding some biota artificially or controlling established, nontarget species enhance colonization.

The increase in management inputs moving from unassisted succession to a technical solution can be viewed as a sequential removal of barriers (filters) to species colonization and persistence (Temperton 2004; see Chapters 2 and 7). In the early stages of spontaneous succession there are a number of physical, chemical, and dispersal barriers to species establishment. As succession proceeds, these barriers are likely to decrease in importance and be replaced by

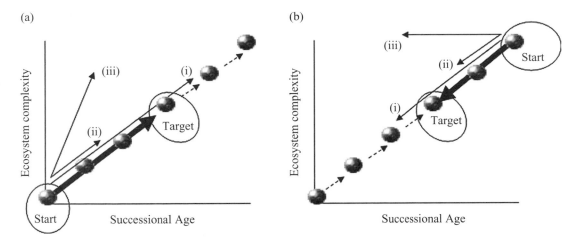

Figure 6.1 Hypothetical diagrams outlining the potential outcomes of management to manipulate succession with an assumption that there is increasing complexity in each successional stage. (a) The usual model of succession where there is increasing complexity with successional age (bold arrows + dotted arrows). Here succession is manipulated to follow this natural process; the ideal management is the bold arrow. (b) Succession reversal, where the succession is manipulated to reverse the process; again the ideal management is the bold arrow. In (a) and (b) other management options are shown: (i) overshoot where the management goes too far, (ii) undershoot where the target is not reached but the direction is correct, and (iii) where the succession deviates from the target trajectory.

barriers imposed by competition from developing vegetation. With a technical solution the aim is to remove some or all of these barriers artificially.

The choice of strategies is, therefore, not easy. For any given restoration scheme there is usually a wide range of options, and the final decision may involve participation of a variety of stakeholders. However, as a general principle we support the idea of using the minimum treatment required (Bradshaw 2002) in order to reduce cost and effort during restoration.

From a generalized viewpoint of ecological succession, there are two principal directions by which succession can be manipulated to attain a target, either to accelerate it or to reverse it if it has proceeded beyond the target (Fig. 6.1). In the former case, the starting capital of soil nutrients and biota is often low or even nonexistent. Here, there is a need to build an ecosystem that will hopefully follow a successional trajectory with increasing ecosystem complexity. The latter case is almost exactly the reverse (Marrs and Bradshaw 1993, Marrs 2002).

We distinguish between two management phases: a *Primary Management Phase* and a *Secondary Management Phase*. The former includes manipulation before succession starts. The latter covers enhancement of desirable (target) species, control and eradication of nontarget species (weeds and aliens), and manipulation of succession at the community level by changing management within an ongoing succession. Many of the methods mentioned below can be used in both phases.

6.3.2 Physicochemical Manipulation

There are a large number of ways that the physical structure of a site can be manipulated (Whisenant 2002). Often a site may be completely reshaped by earth-moving equipment to help it blend into the surrounding landscape. The steepness of slopes can be reduced, and the drainage regime altered to achieve the required moisture conditions. When large-scale earth-moving machinery is used to change site/soil conditions, it can damage the substrate through compaction, especially if the soil is wet. Where this occurs, additional treatments such as surface ripping or drainage may be needed (Montalvo *et al.* 2002).

Chemical manipulations may also need to be done, especially where there is a deficiency or toxicity of chemical elements in the soil-forming material. In many mine wastes, for example, there is almost no organic matter, a very low nutrient supply, and, depending on parent material, a soil pH between 2 and 12. Fertilizers will often need to be added if the soil material is very deficient in major plant nutrients (Bradshaw 1983, Marrs and Bradshaw 1993). Sometimes, adding soil-forming materials such as green wastes, industrial wastes (e.g., paper pulp wastes), or sewage sludge can improve site conditions (Greipsson 2002).

The aim of physical and chemical manipulations is to produce a site that is well contoured into the local landscape and with appropriate drainage or irrigation for the target ecosystem to establish, and where the soil is suitable to allow the required species to establish and grow well.

6.3.3 Biological Manipulation

Adding seeds of the target species to the system is the most common biological manipulation, although it is possible to use other propagules or transplants from donor communities (Hodder and Bullock 1997, Antonsen and Olsson 2005). Once introduced, steps must be taken to ensure their establishment and persistence.

Where seeds are added, they can be treated using germination-promoting agents to ensure a rapid germination. At the same time it is often sensible to add microorganisms such as the appropriate mycorrhizas and *Rhizobium* bacteria if leguminous plants are to be included in the seed mix (Greipsson 2002; see Chapter 3).

One major issue is the provenance of seeds and of plant material derived vegetatively. Over the last 20 years, there has been a large increase in the commercial availability of seeds of seminatural biotopes from native sources, and in some places seeds of species can be obtained from very localized sources (e.g., http://www.floralocale.org). On the other hand, most commercial mixtures or transplants do not respect regionality (McKay *et al.* 2005). In general, when selecting material for sowing or transplanting, we advise that whenever possible, seeds and other propagules should be used from local sources. When transplanted material is to be used, material of a single clone should be avoided, in order to maximize genetic diversity. High genetic diversity may reduce the risk of extinction under fluctuating environments and allow a species to occupy more microhabitats.

In some situations, succession is circumvented by importing an ecosystem from elsewhere, either completely in blocks (transplants) or in part. The most usual case is to import topsoil from elsewhere; where this is done, nutrients, seed banks, and soil microflora can be imported in one operation, which then includes both the physicochemical and biological manipulation (Vécrin and Muller 2003).

6.4 Habitats: Contrasting Problems and Solutions

In this chapter we focus on aspects of ecological restoration in selected habitats, based mainly on our personal experience. However, we believe that the selection of habitats below illustrates the variability and potential for manipulating succession, and that the generic aspects of these case studies have a wider applicability.

6.4.1 Succession in Heavily Disturbed Habitats

6.4.1.1 Mining Sites

Mining activity despoils *ca.* 1% of the land surface (Walker 1999). Technical reclamation has usually prevailed, consisting of rough manipulations of the substrates such as remodeling, drainage, and covering the surface using various organic materials followed by restoration toward either grassland or forest (Whisenant 1999). Unassisted succession has been rarely included intentionally as a part of a restoration project. In temperate Europe, for example, unassisted succession has produced more or less stabilized, seminatural vegetation in 20–30 years (Wolf 1985, Prach 1987, Kirmer and Mahn 2001). In a comparison of unassisted succession and technical reclamation schemes on coal spoil heaps in one of the largest active coal mining districts in Europe (northwestern Czech Republic), sites with unassisted succession had double the number of plant species and fewer invasive species (Fig. 6.2). Moreover, no investment was needed compared to the high costs ($50,000 ha^{-1}) of technically reclaimed sites (Hodačová and Prach 2003). Spoil heaps, which revegetated spontaneously, also provided better refugia for rare and endangered species and did not exhibit any negative off-site effects (Benkewitz *et al.* 2002).

Figure 6.2 A spoil heap from coal mining in northwestern Czech Republic. Unassisted succession led after 30 years to a diverse and more or less stabilized late successional stage that harbored double the number of plant species than technically reclaimed heaps of the same age.

There are, however, situations where unassisted or slightly manipulated succession is not effective; e.g., on very acidic or toxic substrata, or under extremely dry conditions. Acidic sites in an East German coal mining district were without any vegetation 70 years after abandonment (Wiegleb and Felinks 2001). In such cases, physical manipulation of the environment is essential within the *Primary Management Phase*. Liming, topsoiling, and covering of the surface by other organic or inert material are frequently used. Hydroseeding may help to overcome adverse site moisture conditions (Munshower 1994, Ninot *et al.* 2001). Some technical reclamation is also needed on easily eroded sites: in northeastern Spain, unassisted succession created seminatural communities on dumps from coal mining. However, the vegetation cover produced was inadequate for site stability (Ninot *et al.* 2001). Technical measures are also needed in the case of outputs of toxic substances from some mine tailings (Whisenant 1999).

We are convinced that unassisted succession should be suitable for restoring many mine wastes. Unfortunately, this is often prevented by legislation or practice that obligates mining companies to restore disturbed sites quickly. In some countries, there has recently been progress in this area; in some German coal mining districts, at least 15% of the area disturbed by mining must be left for unassisted succession (Wiegleb and Felinks 2001).

Sand and gravel pits and stone quarries are best left for unassisted succession, especially if the disturbed site is small (Prach 2003; Figs. 6.3 and 6.4). These sites provide important nutrient-poor refugia for species under threat in the surrounding fertile landscape. If mining sites are not too extensive, do not produce any pollution, do not damage any valuable locality, and are aesthetically acceptable, they can even increase landscape diversity. These conclusions

Figure 6.3 An abandoned sand pit in southern Czech Republic, 12 years after cessation of sand extraction. A spontaneous stand, dominated by Scots pine (*Pinus sylvestris*) appears at the rear of the photo, planted Scots pine in the foreground. The technical reclamation was not needed as unassisted succession proceeded faster toward the target of a seminatural pine forest.

Figure 6.4 A 50-year-old abandoned limestone quarry in the Bohemian Karst area of central Czech Republic. Unassisted succession proceeded toward seminatural stages represented by species-rich dry grasslands, shrubs, and woodland.

are supported by studies of basalt quarries and sand and gravel pits in the Czech Republic, Central Europe (Novák and Prach 2003, Řehounková and Prach, 2006). Within *ca.* 20 years, unassisted succession in basalt quarries led to seminatural vegetation very similar to natural steppe-like communities, and contained many target species, including endangered ones. The establishment of target species was, however, related to the presence of natural communities within 30 m of the quarry, highlighting the importance of a local species pool and poor dispersal of many target species. In abandoned sand and gravel pits in the same region, seminatural vegetation, ranging from steppe-like vegetation, woodlands, and wetlands, started to establish immediately after abandonment in most cases, and reached more or less stable, late-successional stages within 40 years. In central Europe, the worst option is to use a technical solution for reclamation of the mining sites that includes the use of organic amendments before planting or seeding (Prach and Pyšek 2001, Prach 2003). The increased nutrient supply usually stimulates the establishment and expansion of competitive ruderals which then form vegetation of low ecological interest and inhibit establishment of target species. However, such technical measures may be justified under extreme climatic or substratum conditions where an amelioration is needed.

6.4.1.2 Abandoned Fields

Abandonment of arable land is a worldwide phenomenon (Rejmánek and van Katwyk 2004). In Europe, soil left after agricultural abandonment is usually fertile, although sustained arable use can reduce the organic matter content (Marrs 1993); thus there should be no need for physical manipulation. Only when heavily fertilized soil is present, topsoil removal or sod cutting may be needed to initiate succession toward less productive stages (Verhagen *et al.*

Figure 6.5 Spontaneous succession of woody species (predominantly *Sambucus nigra*) into an abandoned bus. The establishment was enhanced by perching birds and decreased competition from the herb layer inside the bus.

2001). Usually, manipulation of succession in abandoned fields can be viewed as a part of the *Secondary Management Phase*, and used to accelerate succession toward a target ecosystem, e.g., grass expansion for use as hay meadows or pastures (WallisDeVries *et al.* 1998) or establishment of woody species to create woodland (Olsson 1987; Fig. 6.5).

In order to promote grassland development, it is often sufficient to introduce mowing or grazing (Bakker 1989, WallisDeVries *et al.* 1998; see Chapter 5) because grasses usually establish spontaneously, in temperate zones mostly within 15 years (but see Pywell *et al.* 2002a), although this can be enhanced by seeding. We recommend that where seeding is to be implemented it is done in the initial stages, just after abandonment, where the vegetation tends to consist of annual species, and seedling establishment is easier in a less-competitive environment. Later in the succession, competitive perennial forbs usually dominate and colonization is more difficult (Osbornová *et al.* 1990). Unfortunately, commercial seed mixtures, not respecting local species pool, are predominantly used to convert arable land or recently abandoned fields into grassland, and the densely sown vegetation may slow down the establishment of target species (Vécrin *et al.* 2002).

If the grassland is left alone, woody species usually establish. Woodland restoration in former arable land has often used a technical solution (traditional afforestation), and this has resulted in monospecific plantations. There are fewer examples of intentional restoration using unassisted succession of woody species in Europe. The establishment of woody species is heavily influenced by seed availability in the surroundings (Olsson 1987). Site moisture conditions are often critical for the establishment of woody species (Osbornová *et al.* 1990). Establishment can be restricted on dry sites by physiological constraints and on wet sites by competition from robust, productive grasses, and herbs. Under mesic conditions, a dense, shrub woodland develops spontaneously after 20 years in most studies in temperate Europe (Table 6.1). Where shrubs and trees have developed on a site where grassland or heathland is the target,

Table 6.1 Late successional stages that have developed spontaneously on abandoned arable land in temperate Europe.

Dominant species	Age [yr]	Region	Site conditions	Human activity	References
Fraxinus excelsior, Acer campestre, Quercus robur	>30	UK	Mesic, chalk	None	Harmer *et al.* (2001)
Fraxinus excelsior, Salix caprea	>12	C Germany	Mesic, loamy soil	None	Schmidt (1983,1988)
Picea abies, Betula spp.	>30	Aland Isl. (Finland)	Wet, small sites surrounded by forest	None	Prach (1985)
Alnus incana	>20	C Finland	Wet	None	Prach (1985)
Populus tremula, Quercus robur, Tilia cordata	>21	C Sweden	Mesic	None	Olsson (1987)
Pinus sylvestris, Juniperus communis	>25	NE Poland	Sandy, nutrient poor	None	Falinski (1980) Symonides (1985)
Pinus sylvestris	>10	Czech Republic	Sandy, nutrient poor	None	Prach, unpubl.
Salix cinerea	>10	Czech Republic	Wet, moderate in nutrients	None	Prach, unpubl.
Phragmites australis	>30	Czech Republic	Wet alluvial site, rich in nutrients	None	Prach, unpubl.
Crataegus spp.	>25	Czech Republic	Mesic, moderate in nutrients	None	Osbornová *et al.* (1990)
Festuca rupicola	>30	Czech Republic	Dry, moderate in nutrients	None	Osbornová *et al.* (1990)
Festuca rupicola	>12	C Romania	Dry, nutrient poor	Extensive grazing	Ruprecht (2005, 2006)
Festuca vaginata, Stipa borysthenica	>24	C Hungary	Dry, sandy, nutrient poor	Sheep grazing	Csecserits & Rédei (2001)
Robinia pseudoacacia	>24	C Hungary	Dry, sandy, nutrient poor	None	Csecserits & Rédei (2001)

successional reversal is needed (see Section 6.3). Here, cutting and subsequent herbicide application might be useful (Marrs 1988).

Unassisted succession on abandoned fields in temperate Europe has often been very successful, leading to seminatural vegetation with high nature conservation value (Table 6.1; see Chapter 5). This approach is most successful in landscapes that have retained traditional agricultural practices, where soils have not been altered markedly by over-fertilization or drainage, and where natural habitats still frequently occur in the vicinity (Falinski 1980, Ruprecht 2005). Conversely, in altered landscapes, competitive ruderals (generalists) or aliens may expand (Prach *et al.* in press; see Chapter 3). This was demonstrated on ungrazed, abandoned fields on sandy soils in Hungary, where invasion by the alien *Robinia pseudoacacia* was problematic (Csecserits and Rédei 2001).

It is difficult to produce accurate predictions of likely outcomes of unassisted oldfield succession due to its high stochasticity (Rejmánek 1990). Site history, moisture status, soil fertility, character of the surrounding landscape, management implemented, interactions between trophic levels, and random effects interact and influence the eventual outcome (Pickett *et al.* 1987, Tilman 1988). Despite these limitations, robust predictions of successional trajectories at a country-wide scale and their use in restoration programs were possible after simple categorization of fields on the basis of moisture (wet, mesic, dry) and nutrient supply (rich, intermediate, nutrient-poor) (Prach *et al.* 1999).

6.4.2 Succession in Less Disturbed Ecosystems

In the past, grasslands, heathlands, and various wetlands have been created as a result of human activity (Gimingham 1992). In Europe, these are important seminatural ecosystems from a conservation viewpoint. They have substantially degraded and decreased in their extent, but still contain many rare and retreating species. Their restoration is of high priority.

6.4.2.1 Temperate Grasslands

Species-rich grasslands have declined substantially in Europe because of agricultural developments during the latter half of the 20th century. They were used earlier for hay-making and low-intensity grazing (Bakker 1989). Grasslands on very infertile soils were abandoned whereas on the more fertile soils exploitation was intensified, including conversion to arable land. Abandonment led to a decrease in species-richness and a loss of species characteristic of grassland communities (Diemer *et al.* 2001, Matějková *et al.* 2003). The development of cheap fertilizers enabled intensification and increased soil fertility and crop productivity, and changed the species composition from slow-growing, small species to fast-growing, competitive, tall species. As these productive fields could be used earlier in the season, and more intensively, the competitive advantage shifted toward early-flowering species, and especially clonal perennials (van Diggelen *et al.* 2005).

The normal practice for ecological restoration of grasslands in Europe is to reintroduce a mowing or grazing regime without adding fertilizer. In the case where grasslands have been abandoned this may work (Hansson and Fogelfors 2000), but not always (Matějková *et al.* 2003). In intensively exploited grasslands this approach often takes a long time (Oomes and Mooi 1981, Bakker *et al.* 2002) because large amounts of nutrients have accumulated in the soil. Nutrient balance sheets have shown that the annual removal of N and P through haymaking is at most 3% (N) to 5% (P) of the soil nutrient pool (Bakker 1989), and the amounts removed by grazing are very low (Perkins 1978, Bakker 1989).

One way to decrease nutrient supplies rapidly is to completely remove the surface layer of topsoil (Marrs 1993, Verhagen *et al.* 2001). This approach has been shown to reduce soil N, but it was less effective at reducing soil P (Marrs *et al.* 1998b). This approach may also lead to a shift in the nutrients which limit productivity; from co-limitation by N and P to limitation by N alone. The consequences of such a shift are unknown but indications suggest that it might favor grasses at the expense of herbs (Olde Venterink *et al.* 2003).

On organic-rich grassland soils the nutrient stocks are large, and the only way to reduce fertility is to lower nutrient turnover rates. One way to lower the N supply is to reduce oxygen availability by rewetting drained sites to lower aerobic microbial activity (see Chapter 5). This may work in sites that have not been drained intensively, but where drainage has been intense the N supply remains high (Hauschild and Scheffer 1995). This is also generally the case immediately after rewetting. Phosphorus availability is mainly controlled by the adsorption and desorption of P on Fe-, Al- and Ca-complexes. Alteration of these binding reactions when sites are rewetted with mineral-poor rainwater or sulphate-rich water may even lead to increased P supply (Lucassen *et al.* 2004). Other impacts are also possible; infiltrating rainwater can wash out potassium (K) and K-limitation may develop (Eschner and Liste 1995, van Duren *et al.* 1997) instead of N- and/or P-limitation. Again, the consequences of such shifts are unknown and manipulation of these factors is not easy.

The local species pool (cf. Zobel *et al.* 1998) is a second filter that affects succession trajectories of grassland restoration. There are negative relationships between the period of intensive land use and the degree to which species reappear from the seedbank (Bekker *et al.* 1997), and between the degree of landscape fragmentation and immigration of target species (Poschlod and Bonn 1998). Thus, the speed and course of succession are determined largely by the period and intensity of previous land use on the site and in the surroundings.

Because of limitations on species colonization, steps need to be taken to increase the colonization of target species. Successful methods included deliberate mowing or grazing regimes where the machines or animals are moved from species-rich target communities to species-poor restoration sites in order to facilitate the transfer of seeds (Strykstra *et al.* 1997, Mouissie *et al.* 2006). Alternatively, dispersal limitation can be overcome by deliberate addition of propagules by a variety of methods; for example, by adding fresh hay cut from species-rich reference fields (Hölzel and Otte 2003), adding top soil from donor sites (Vécrin and Muller 2003), or even by transfer of complete turfs (Klötzli 1987, Šeffer and Stanová 1999, Antonsen and Olsson 2005). The obvious disadvantages of these methods are that reference donor sites are affected severely or even destroyed and these methods are very costly (Vécrin and Mueller 2003).

Unassisted succession in human-managed secondary grasslands (in contrast, for example, to natural prairies) is not usually a tool of restoration. Continued manipulation of succession in the form of maintenance management (Bakker and Londo 1998) is essential for the persistence of seminatural grasslands.

Table 6.2 Situations where heath communities have been restored.

Situation	Problems	Is unassisted succession possible?	Restoration methods (in approximate order used)	Selected references
Raw mineral wastes	Lack of seeds and nutrients	Yes—but takes a long time (25 years)	1. Regrade site, add organic amendments if possible 2. Add seed 3. Add fertilizerr 4. Use nurse crops 5. Transplants—plants or turfs 6. Grazing control	Roberts *et al.* (1981), Anonymous (1988)
Agricultural land	Lack of seeds, growth of ruderal species, pH too high	No, soil conditions often changed during agriculture	1. Acidify soils 2. Reduce fertility, (e.g., by sod cutting) 3. Grazing control 4. Add seeds	Pywell *et al* (1994, 1997), Davy *et al.* (1998), Dunsford *et al.* (1998), Owen and Marrs (2000a,b)
Succession reversal	Expansion of late-successional species e.g., *Betula* spp., *Pinus sylvestris*, *Rhododendron ponticum*, *Pteridium aquilinum*, *Deschampsia flexuosa, Molinia caerulea*	No, except insofar as heathland may exist in a temporary stage if the late-successional stage is damaged	1. Reduce fertility 2. Control late-successional species: mechanical or herbicidal methods 3. Create conditions for heathland species to germinate 4. Add seeds 5. Grazing control	Marrs (1988), Milligan *et al.* (2004)

6.4.2.2 Heathlands

Restoration of heathland communities has been described in a number of situations, from establishment on raw mineral wastes (primary succession), from abandoned arable land (secondary succession), and where succession has occurred through inappropriate management of existing heathlands (Table 6.2). For heathlands, unassisted succession is a serious possibility for restoration of raw substrates (e.g., sand and mining wastes) where there is sufficient seed rain from surrounding areas (Roberts *et al.* 1981). On sand wastes, succession was slow (*ca.* 40–60 years for a heathland to develop) and highly variable (Roberts *et al.* 1981). Technical solutions can accelerate this process and provide a greater vegetation cover.

In raw wastes and agricultural land, seeds of heathland species (usually *Calluna vulgaris*) are scarce and may even be absent (Pywell *et al.* 1997), whereas in late-successional stands invaded by woodland *Calluna,* seed banks can persist for greater than 70 years (Pywell *et al.* 2002b). Seed limitations can be overcome by adding seeds, shoots with the attached seed capsules, or topsoil. Often topsoil use provides a more diverse flora than seeds alone because the soil contains a greater species diversity (Pywell *et al.* 1994). It is also possible to either use transplanted turfs or nursery-grown plants (Webb 2002). Where topsoil or turfs are used there are implications for the donor site and cost.

To restore abandoned land there are two potential constraints, high pH and high nutrient supplies (usually P), both linked to past fertilizer/lime additions. There are a variety of techniques available for reducing soil pH such as the addition of elemental or pelleted S or addition of acidic plant materials. The most effective has been the addition of S, either directly (Owen *et al.* 1999), or as pyrite-rich peat (Davy *et al.* 1998). Experience has shown that application rates need to be calculated empirically for individual sites (Owen *et al.* 1999).

When attempting to set back succession on heathlands where succession has proceeded to woodland stages, it is essential to remove the woodland species (in Europe these are usually *Betula* spp., *Pinus sylvestris*, *Pteridium aquilinum*, *Rhododendron ponticum*) and especially their litter. Control of the colonizing species is essential (Marrs *et al.* 1998a). Control of conifers such as *P. sylvestris* is done easily by cutting, because it does not regenerate from cut stumps, but for all other colonizing species, either mechanical treatment needs to be repeated on a regular basis (e.g., *Pteridium aquilinum* control; Marrs *et al.* 1998a), or a herbicide should be included in the strategy (Marrs 1988). One of the techniques used in heathlands is "sod cutting and removal," and subsequent restoration proceeds in a nutrient-poor, subsurface layer (Werger *et al.* 1985, Diemont 1994). This approach has been commonly used in The Netherlands and Belgium.

Like grasslands, heathlands require manipulation in the *Secondary Management Phase*, and usually this will include grazing, cutting, and burning. However, young heathland plants are very sensitive to grazing, and they need protection from grazing until they become established (Gimingham 1992).

6.4.2.3 Wetlands

Highly dynamic systems such as river floodplains have high nutrient turnover rates and are normally very productive. In Europe, such systems were once dominated by species such as *Carex* spp., *Phragmites australis*, and *Typha* spp. in the wettest parts and soft- and hardwood forests in less-flooded parts. Because of their fertility, the majority of such sites are exploited for agriculture and pristine floodplains are now amongst the most endangered ecosystems worldwide (Olson and Dinerstein 1998). After abandonment, restoration to sedge or reed beds and softwood forests (e.g., *Salix* spp., *Populus* spp.) can proceed quickly but a conversion to hardwood forests (e.g., *Quercus* spp., *Ulmus* spp., *Frangula* spp.) is less common. Existing evidence of hardwood development comes mainly from North America and results suggest that unassisted succession takes at least 50–100 years, depending on site conditions and proximity to seed sources (Collins and Montgomery 2002). Planting native tree mixtures can accelerate this process somewhat but even after 50 years there were considerable differences in the understory composition of planted woodland and historic bottomland forests (Shear *et al.* 1996).

Wet systems with less dynamic water regimes (see Chapter 5) usually exhibit lower site fertility, because organic matter decomposes slowly and is deposited as peat. The nutrient stocks of such mires are large but nutrient availability is low (Koerselman and Verhoeven 1995). Human interference almost always involves drainage, with a consequent initial large release of stored nutrients (Grootjans *et al.* 1985). Even after just a short drainage period and soil structural change, nutrient availability increases, remaining high even when such sites are rewetted (Eschner and Liste 1995). It is impossible to predict whether such systems will redevelop into nutrient-poor mires or develop into a more productive alternative stable state (cf. Scheffer 1990). However, where the altered topsoil has been removed to expose nondegraded layers and the system rewetted, it has been possible to restore target communities with low-productivity (Pfadenhauer *et al.* 2001, van der Hoek 2005).

Restoration is particularly difficult on large-scale, surface-mined *Sphagnum* bogs, even when unaltered peat layers are left (Money 1995). The most likely reason for this lack of success is the inability to maintain stable water levels in leftover remnants (Giller and Wheeler 1988, Joosten 1993). If stable water levels can be achieved, bog succession can be rapid, taking only 10–20 years on floating rafts that fluctuate with the water table (van Diggelen *et al.* 1996, Beltman *et al.* 1996b). The intensity of restoration management required depends on the damage inflicted on the system. Low-quality woodland (*Pinus sylvestris*, *Picea abies*, *Betula* spp.) usually develops after large-scale industrial peat harvesting unless the water table is manipulated (Salonen *et al.* 1992, Prach and Pyšek 2001). Where peat has been extracted in a traditional manner (i.e., shallow and without deep drainage), unassisted succession is usually successful (Joosten 1993).

Base-rich fens are also affected by extraneous factors, especially long-distance hydrological interference that redirects groundwater flow patterns and results in changed hydrochemical conditions. A decrease in upwelling, base-rich groundwater normally leads to acidification of the top soil, and also to increased water level fluctuations. Succession will not, therefore, lead to the reestablishment of low-production fen communities (Wolejko et al. 1994, Wassen *et al.* 1996, van Diggelen and Grootjans 1999), unless the previous hydrological system is restored. If this is not the case there will be very rapid succession toward bogs, even in calcareous landscapes (Jasnowski and Kowalski 1978). In contrast, the life span of calciphilous pioneer communities in dune slacks can be extended for many decades in a Ca-poor landscape by upwelling, Ca-rich groundwater (Lammerts and Grootjans 1998).

A second situation where base-rich fen vegetation develops quickly without much active intervention is in former peat cuttings which have become filled with base-rich ground and surface water. As long as the rafts are still thin, the pH remains high enough for basophilous species, but after it has achieved a certain thickness it becomes isolated from the underlying water body and pH starts to decrease. In medium-sized turf ponds (1 ha) this phase is normally reached within a few decades (van Wirdum 1995, van Diggelen *et al.* 1996). This phase can be prolonged if rainwater is removed artificially by a shallow drainage system (Beltman *et al.* 1995), or the process can be started by new cuttings (Beltman *et al.* 1996a).

Salt marshes are an even more extreme type of wetland ecosystem with two major constraints for plant growth: (1) low oxygen availability, and (2) high salt concentration. Only a few species are adapted to these extremes but all are very characteristic and are mostly restricted to this habitat. The majority of salt marshes in the temperate zone have been managed (Bakker *et al.* 1997). Mature salt marshes used to be artificially enclosed with banks for grazing but over the last 25 years this management has reduced and grazing has diminished (Esselink 2000). The consequence is a reduction in salt-marsh pioneer phases and a domination by late-successional species; e.g., *Elymus athericus* (Bos *et al.* 2002) or *Phragmites australis* (Esselink *et al.* 2000). The overall result is a loss in biodiversity. Succession in salt marshes can be manipulated back to earlier stages by reopening existing embankments and flooding. In a survey of 70 restored flooded sites in northwestern Europe, the percentage of target species was as high as 70% in the best examples (Wolters *et al.* 2005).

6.5 The Role of Aliens

6.5.1 Successional Pattern: Repairing Function of Native Vegetation

Recent attempts to bring the science of invasive species and succession together have proved profitable (Davis *et al.* 2000, 2005; see Chapter 3). Alien species (defined as those whose presence in a given area is due to intentional or unintentional human involvement, or which have arrived without the help of people from an area in which they are alien; see Pyšek *et al.* 2004b) are increasingly common in successional seres. For example, in 55 successional studies in central Europe and North America, 25% of the species (range 2–81%) were aliens. Aliens were most prevalent in ruderal habitats and old fields, and their representation declined during the successional process. The rate of this decrease was context-dependent: industrial habitats had a greater proportion of aliens at the start, but a faster decline with time than habitats associated with agricultural landscapes. Alien species contributed more in terms of species number than cover reflecting that many of them are rare casuals (Richardson *et al.* 2000, Pyšek *et al.* 2004b). Those aliens classified as neophytes (species introduced after 1500AD) were most likely to become dominants (Pyšek *et al.* 2004a).

It is not known how much the pattern of decrease of alien species during succession is determined by the exposure of various successional stages to different propagule pressures of alien species. It is predicted that colonization by diaspores will be greater at the beginning of succession (Rejmánek 1989). Experimental studies on the invasibility of successional stages are rare, but Bastl *et al.* (1997) found that early, but not initial, successional stages were most prone to plant invasions. The establishment of aliens in the initial stages of succession was probably restricted by adverse abiotic conditions, whereas in later successional stages, intensive competition from resident species appeared more important. This seems to be a common pattern (Rejmánek 1989) and should be considered in restoration practice. It has been suggested that the maximum cover and proportion of aliens are found in the initial stages of mesic succession (Rejmánek 1989). These results suggest support for the successional repairing function of native vegetation (Rejmánek 1989) indicating that during spontaneous succession alien species should disappear in time (Pyšek *et al.* 2004a).

6.5.2 Manipulating Succession in Invaded Sites

The unassisted recovery of native vegetation during succession is of little practical use when dealing with those large-scale invasions where immediate action is needed. In such cases, the dominant alien species must be at least contained and preferably controlled or eradicated (Myers and Bazely 2003). However, species removal in isolation can result in unexpected changes to other ecosystem components, such as trophic interactions (Zavaleta *et al.* 2001), and open the way to reinvasion by the problem weed species or other aliens. To avoid this reinvasion, appropriate restoration measures need to be taken, and this is analogous to removal of late-successional species when succession is being reversed, e.g., on heathlands (see Sections 6.3.1. and 6.4.2b).

With increasing dominance of an invasive species, and increasing difficulty to control it, possible action moves along the sequence from unassisted succession to biological manipulation to technical reclamation. Planting or sowing indigenous species should accelerate recovery of resident vegetation in highly degraded sites after the clearance of dense and extensive stands of alien plants, and where the likelihood of recovery from seed banks is low (Richardson *et al.* in press). Appropriate management, aimed at manipulating the site characteristics is difficult. Moreover, some highly invasive species (transformers *sensu* Richardson *et al.* 2000) alter ecosystem functioning and change the site conditions.

Conventional wisdom suggests that disturbance, and both site nutrient and moisture status as reflected in site productivity, all play a crucial role in plant invasions (Rejmánek 1989, Hobbs and Huenneke 1992, Davis *et al.* 2000). Manipulation of both the disturbance regime and productivity are, therefore, two major options for the subsequent control of alien species. Manipulation of disturbance is often aimed at restoring a management regime that was typically applied to the habitat before invasion. The effect of such management depends on the site productivity. In productive environments a reduction in disturbance is likely to allow native species to outcompete aliens (Huston 2004). However, manipulating productivity is much more difficult to manage than disturbance.

Because invasions are context-dependent and individual habitats differ largely in the level of invasion and invasibility (Lonsdale 1999, Chytrý *et al.* 2005), it is essential that the character of the restored site is taken into account. Spontaneous succession either assisted or not, may be less efficient in riparian zones because of their high water-flow dynamics and continuous addition of propagules of alien species. The nature and effects of the fluctuating conditions, such as timing of floods, make the role of aliens in riparian succession difficult to predict, especially on recolonization processes after removal. The restoration strategy must, therefore, take into account the pattern of arrival of alien species propagules into a site subjected to restoration (Richardson *et al.* in press). Clearly, the source of reinvasion is likely to be from upstream, so if aliens are to be controlled then upstream source populations must be identified and controlled before restoration is started. The pattern of propagule arrival is much more difficult to predict in habitats outside river corridors where the arrival of propagules is nondirectional, from various sources and highly stochastic. Where the pool of alien species is large and diverse (e.g., urban wasteland) sowing or planting of native species may prevent invasion.

Under certain circumstances, aliens are used intentionally in restoration programs, especially where economic priorities prevail (see Chapter 5). For example, in dry areas of Africa, alien woody species (*Albizia* spp., *Acacia* spp., *Eucalyptus* spp., or *Prosopis* spp.) are largely planted into different successional stages to increase productivity and help prevent desertification (Adams

2002), in addition to their dual purposes of fiber production and economic gain. Under these circumstances it is perhaps justified. In temperate Europe, alien woody species are unfortunately still used, especially in the technical reclamation of derelict sites. The species used might even include ones recognized as highly invasive, e.g., *Acer negundo, Quercus rubra, Robinia pseudoacacia*, and *Pinus strobus*. We strongly discourage the use of such aliens in restoration programs because of their possible uncontrolled subsequent spread. Moreover, alien woody species can usually be replaced by native taxa providing comparable economic profit.

6.6 Implementation of Scientific Knowledge into Restoration Plans

The major task for scientists engaged in restoration programs is to advise on the biotic/abiotic state of the ecosystem (the boundary conditions of restoration) and to suggest pathways that are likely to occur on a restored site whether it has been exposed to unassisted succession or active manipulation. Restoration ecologists should be able to specify the most suitable target communities (Perrow and Davy 2002) from a possibly extensive list of options and compare their predictions with reality through monitoring. To achieve a successful prediction, information from three sources can be exploited: (i) detailed quantitative case studies; (ii) comparative studies over a larger geographical range and across environmental gradients; and (iii) qualitative and site specific knowledge based on local information (Prach *et al.* 2001). The best scenario is obviously where results are available from a detailed case study from a site under restoration, or if there is time and resources, to schedule and carry out a pilot study. This situation is not common and the pilot study is not always carried out in the same conditions of soil type and intensity of disturbance. Therefore, predictions are often based on less precise information.

Emphasis is often placed on the low predictability and high stochasticity of succession, especially at the species level (Pickett *et al.* 2001, Fukami *et al.* 2005). We are less skeptical and consider both unassisted and manipulated successional trajectories to be predictable to a certain degree required for projected restoration schemes. The level may only represent growth forms or functional species groups rather than a detailed species sequence. Report cards based on indicators of abiotic and species changes can also be used to guide restoration (Walker and Reuter 1996).

Various Expert Systems or Decision-Support-Systems can be developed to transfer the knowledge of succession and its manipulation into restoration programs; unfortunately, few have been published or used (Hill 1990, Hunt *et al.* 1991, Prach *et al.* 1999, Hill *et al.* 2005). Such systems should be easy to use and provide straightforward, robust answers to simple questions (Luken 1990). For practical use, they are more efficient than mathematical models because they are based on a wider variety of information and not only on quantitative data and mathematically derived functions. Vague, intuitive knowledge and precise quantitative information can be successfully combined (Noble 1987). Using this approach, the expert system SUCCESS predicts the sequence of spontaneous seral stages and dominant species change in various human-disturbed habitats in central Europe (Prach *et al.* 1999). The system has a potential to be extended and include predictions of the manipulation of succession.

We emphasize the role of monitoring in evaluating the success of manipulation measures against predeterminend restoration targets. Monitoring provides a feedback by which the restoration program can be modified, and at the same time deliver information to improve our knowledge of succession.

6.7 Conclusions

Despite some recent progress, unassisted succession and ecologically sound manipulation of spontaneous succession as a part of restoration projects are exploited less often than they could be. Technical reclamation using engineering or mechanical approaches still dominate many restoration projects. In some cases, we can rely on unassisted succession, which can provide better and cheaper results than technical reclamation (see Section 6.4.1). That unassisted succession can take longer to reach the target than technical reclamation is compensated by the higher structural and functional diversity and higher natural and conservation value of resulting vegetation.

Whether to use technical reclamation or spontaneous succession, manipulated or not, may depend on the position of the disturbed site on the productivity gradient (Fig. 6.6). Numerous case studies (see Section 6.4) indicate that technical reclamation, usually represented by strong physical manipulation of a site, is required most often when site conditions are extreme rather than moderate. At intermediate productivity values, unassisted succession plays a larger role. Unassisted succession is effective especially if a disturbed site is small and surrounded by natural vegetation. On the other hand, unassisted succession itself is not usually a tool of restoration in human managed secondary habitats (see Section 6.4.2). Continued manipulation of succession in the form of maintenance management (Bakker and Londo 1998) is essential for the persistence of the preferred habitats, or succession can be manipulated back to earlier stages.

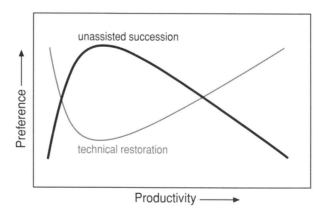

Figure 6.6 Preference for unassisted succession and technical restoration changing along a hypothetical productivity gradient. Unassisted succession is expected to be the best tool of restoration of moderately nutrient-poor sites (e.g., in stone quarries and sand pits), where highly competitive species do not expand. While in extremely unproductive (e.g., toxic) or highly productive (eutrophicated) sites technical restoration may be preferred, either to ameliorate adverse abiotic site conditions or to suppress strong competitors, respectively.

Table 6.3 Theoretical concepts concerning succession and possible corresponding restoration measures to manipulate succession.

Theoretical concepts	Manipulation measures
Primary succession	Physical manipulation
Secondary succession	Biological manipulation, changes in management
Facilitation	Nursery plants
Inhibition	Control or eradication of undesirable competitive species (mechanical, chemical, herbivores, pathogens)
Initial floristic composition	No action
Gap dynamics	Mechanical creation of artificial gaps
Patch dynamics	Rotational management
Intermediate disturbance hypothesis	Adjustment of management
Safe-sites	Physical manipulation, artificial disturbance, mulching
Species-pool	Seeding, planting, eradication of undesirable species in the surroundings, protection of desirable species or communities in the surroundings
Metapopulation theory	Increasing connectivity among restored sites (establishment of corridors, movement of domestic animals, etc.)
Theory of island biogeography	Manipulation of size of a restored site, increasing site heterogeneity

Each restored site requires specific methods and their proper timing to manipulate succession toward a desired target. If the desired target is too specific, then generalizations are difficult. It must be emphasized again that successional processes are highly stochastic and influenced by many factors. Succession may proceed via multiple pathways (Mitchell *et al.* 2000). Because of this complexity, all manipulations of succession must be justified by scientific knowledge. Table 6.3 summarizes theoretical concepts related to succession with possible restoration measures. Effective restoration strategies are best developed with a combination of the succession concepts and local knowledge.

Acknowledgments: We thank Jan Bakker, Anke Jentsch, Lawrence Walker and two anonymous reviewers for their helpful comments on earlier versions of this text. Karel Prach and Petr Pyšek were supported by long-term institutional research plans (AV0Z60050516, Academy of Sciences of the Czech Republic; MSM 0021620828 and 6007665801, Ministry of Education of the Czech Republic); Rob Marrs has received long-term support from NERC and Defra in the United Kingdom. Rudy van Diggelen was supported by grants from the Prince Bernhard Cultural fund in The Netherlands. We thank Dick Visser, Ivan Ostrý, and Zuzana Sixtová for technical support.

References

Adams, W. M. 2002. Restoration policy and infrastructure. Africa. In: *Handbook of Ecological Restoration. Vol. 2. Restoration in Practice*. M. R. Perrow, and A. J. Davy (eds.). Cambridge: Cambridge University Press, pp. 57–77.

Anonymous. 1988. *Heathland Restoration: A Handbook of Techniques*. Southampton: British Gas.

Antonsen, H., and Olsson, P. A. 2005. Relative importance of burning, mowing and species translocation in the restoration of a former boreal hayfield: responses of plant diversity and the microbial community. *Journal of Applied Ecology* 42:337–347.

Bakker, J. P. 1989. *Nature Management by Grazing and Cutting*. Dordrecht: Kluwer Academic Publishers.

Bakker, J. P., Esselink, P., van der Wal, R., and Dijkema, K. S. 1997. Options for restoration and management of coastal salt marshes in Europe. In: *Restoration Ecology and Sustainable Development*. K. M. Urbanska, N. R. Webb, and P. J. Edwards (eds.). Cambridge: Cambridge University Press, pp. 286–322.

Bakker, J. P., Elzinga, J. A., and de Vries, Y. 2002. Effects of long-term cutting in a grassland system: Perspectives for restoration of plant communities on nutrient-poor soils. *Applied Vegetation Science* 5:107–120.

Bakker, J. P., and Londo, G. 1998. Grazing for conservation management in historical perspective. In: *Grazing and Conservation Management*. M. F. Wallis De Vries, J. P. Bakker, and S. E. van Wieren (eds.). Dordrecht: Kluwer Academic Publishers, pp. 23–54.

Bartha, S., Meiners, S. J., Pickett, S. T. A., and Cadenasso, M. L. 2003. Plant colonization windows in a mesic old field succession. *Applied Vegetation Science* 6:205–212.

Bastl, M., Kočár, P., Prach, K., and Pyšek, P. 1997. The effect of successional age and disturbance on the establishment of alien plants in man-made sites: an experimental approach. In: *Plant Invasions: Studies from North America and Europe*. J. H. Brock, M. Wade, P. Pyšek, and D. Green (eds.). Leiden, The Netherlands: Backhuys Publishers, pp. 191–201.

Bekker, R. M., Verweij, G. L., Smith, R. E. N., Reine, R., Bakker, J. P., and Schneider, S. 1997. Soil seed banks in European grasslands: Does land use affect regeneration perspectives? *Journal of Applied Ecology* 34:1293–1310.

Beltman, B., van den Broek, T., and Bloemen, S. 1995. Restoration of acidified rich-fen ecosystems in the Vechtplassen area: Successes and failures. In: *Restoration of Temperate Wetlands*. B. D. Wheeler, S. C. Shaw, W. J. Fojt, and R. A. Robertson (eds.). Chichester: Wiley, pp. 273–286.

Beltman, B., van den Broek, T., Bloemen, S., and Witsel, C. 1996a. Effects of restoration measures on nutrient availability in a formerly nutrient-poor floating fen after acidification and eutrophication. *Biological Conservation* 78:271–277.

Beltman, B., van den Broek, T., van Maanen, K., and Vaneveld, K. 1996b. Measures to develop a rich-fen wetland landscape with a full range of successional states. *Ecological Engineering* 7:299–313.

Benkewitz, S., Tischew, S., and Lebender, A. 2002. "Arche Noah" für Pflanzen? Zur Bedeutung von Altwaldresten für die Wiederbesiedlungsprozesse im Tagebaugebiet Goitsche. *Hercynia N.F.* 35:181–214.

Bos, D., Bakker, J. P., de Vries, Y., and van Lieshout, S. 2002. Long-term vegetation changes in experimentally grazed and ungrazed back-barrier marshes in the Wadden Sea. *Applied Vegetation Science* 5:45–54.

Bradshaw, A. D. 1983. The reconstruction of ecosystems. *Journal of Applied Ecology* 20:1–17.

Bradshaw, A. D. 2002. Using natural processes. In: *The Restoration and Management of Derelict Land: Modern Approaches*. M. H. Wong and A. D. Bradshaw (eds.). New Jersey, London, Singapore, Hong Kong: World Scientific, pp. 181–189.

Christensen, N. L., and Peet, R. K. 1984. Convergence during secondary forest succession. *Journal of Ecology* 72:26–35.

Chytrý, M., Pyšek, P., Tichý, L., Knollová, I., and Danihelka, J. 2005. Invasions by alien plants in the Czech Republic: A quantitative assessment across habitats. *Preslia* 77:339–354.

Collins, B. D., and Montgomery, D. R. 2002. Forest development, wood jams, and restoration of floodplain rivers in the Puget Lowland, Washington. *Restoration Ecology* 10:237–247.

Csecserits, A., and Rédei, T. 2001. Secondary succession on sandy old-fields in Hungary. *Applied Vegetation Science* 4:63–74.

Davis, M. A., Grime, J. P., and Thompson, K. 2000. Fluctuating resources in plant communities: A general theory of invasibility. *Journal of Ecology* 88:528–534.

Davis, M. A., Pergl, J., Truscott, A.-M., Kollman, J., Bakker, J. P., Domenech, R., Prach, K., Prieur-Richard, A.-H., Veeneklaas, R. M., Pyšek, P., del Moral, R., Hobbs, R. J., Collins, S. L., Pickett, S. T. A., and Reich, P. B. 2005. Vegetation change: A reunifying concept in plant ecology. *Perspectives in Plant Ecology, Evolution and Systematics* 7:69–76.

Davy, A. J., Dunsford, S. J., and Free, A. J. 1998. Acidifying peat as an aid to the reconstruction of lowland heath on arable soil: Lysimeter experiments. *Journal of Applied Ecology* 35:649–659.

Diemer, M. W., Oetiker, K., and Billeter, R. 2001. Abandonment alters community composition and canopy structure of Swiss calcareous fens. *Applied Vegetation Science* 4:237–246.

Diemont, W. H. 1994. Effects of removal of organic matter on productivity in heathlands. *Journal of Vegetation Science* 5:409–414.

Dunsford, S. J, Free. A. J., and Davy, A. J. 1998. Acidifying peat as an aid to the reconstruction of lowland heaths on arable soil: A field experiment. *Journal of Applied Ecology* 35:660–672.

Eschner, D., and Liste, H. H. 1995. Stoffdynamik wieder zu vernässende Niedermoore. *Z.f. Kulturtechnik und Landentwicklung* 36:113–116.

Esselink, P. 2000. Nature management of coastal salt marshes: interactions between anthropogenic influences and natural dynamics. PhD-thesis, University of Groningen.

Esselink, P., Zijlstra, W., Dijkema, K. S., and van Diggelen, R. 2000. The effects of decreased management on plant species distribution patterns in a salt marsh nature reserve in the Wadden Sea. *Biological Conservation* 93:61–76.

Falinski, J. B. 1980. Vegetation dynamics and sex structure of the populations of pioneer dioecious woody plants. *Vegetatio* 43:23–38.

Fukami, T., Bezemer, T. M., Mortimer, S. R., and van der Putten, W. H. 2005. Species divergence and trait convergence in experimental plant community assembly. *Ecology Letters* 8:1283–1290.

Gibson, C. W. D., Watt, T. A., and Brown, V. K. 1987. The use of sheep grazing to create species rich grassland from abandoned arable land. *Biological Conservation* 42:165–183.

Giller, K. E., and Wheeler, B. D. 1988. Acidification and succession in a flood-plain mire in the Norfolk Broadland, UK. *Journal of Ecology* 76:849–866.

Gimingham, C. H. 1992. *The Lowland Heathland Management Handbook*. Peterborough: English Nature.

Greipsson, S. 2002. Coastal dunes. In: *Handbook of Ecological Restoration*. Vol. 2. M. R. Perrow and A. J. Davy (eds.). Cambridge: Cambridge University Press, pp. 214–237.

Grime, J. P. 1979. *Plant Strategies and Vegetation Processes*. Chichester: Wiley.

Grootjans, A. P., Schipper, P. C., and van der Windt, H. J. 1985. Influence of drainage on N-mineralization and vegetation response in wet meadows. I: Calthion palustris stands. *Oecologia Plantarum* 6:403–417.

Hansson, M., and Fogelfors, H. 2000. Management of a seminatural grassland; results from a 15-year-old experiment in southern Sweden. *Journal of Vegetation Science* 11:31–38.

Harmer, R., Peterken, G., Kerr, G., and Poulton, P. 2001. Vegetation changes during 100 years of development of two secondary woodlands on abandoned arable land. *Biological Conservation* 101:291–304.

Hauschild, J., and Scheffer, B., 1995. Zur Nitratbildung in Niedermoorböden in Abhängigkeit der Bodenfeuchte (Brutversuche). *Zeitschrift für Kulturtechnik und Landentwicklung* 36:151–152.

Hill, M. O. 1990. Environmental consequences of set-aside land. Report of the Institute of Terrestrial Ecology. Huntingdon, United Kingdom.

Hill, M. J., Braaten, R., Veitch, S. M., Lees, B. G., and Sharma, S. 2005. Multi-criteria decision analysis in spatial decision support: The ASSESS analytical hierarchy process and the role of quantitative methods and spatial explicit analysis. *Environmental Modelling and Software* 20:955–975.

Hobbs, R. J., and Huenneke, L. F. 1992. Disturbance, diversity and invasion: Implications for conservation. *Conservation Biology* 6:324–337.

Hodačová, D., and Prach, K. 2003. Spoil heaps from brown coal mining: technical reclamation vs. spontaneous re-vegetation. *Restoration Ecology* 11:385–391.

Hodder, K. H., and Bullock, J. M. 1997. Translocations of native species in the UK: Implications for biodiversity. *Journal of Applied Ecology* 34:547–565.

Holl, K. D. 2002. Long-term vegetation recovery on reclaimed coal surface mines in the eastern USA. *Journal of Applied Ecology* 39:960–973.

Hölzel, N., and Otte, A. 2003. Restoration of a species-rich flood meadow by topsoil removal and diaspore transfer with plant material. *Applied Vegetation Science* 6:131–140.

Hunt, R., Middleton, D. A. J., Grime, J. P., and Hodgson, J. G. 1991. TRISTAR: An expert system for vegetation process. *Expert Systems* 8:219–226.

Huston, M. A. 2004. Management strategies for plant invasions: Manipulating productivity, disturbance, and competition. *Diversity and Distributions* 10:167–178.

Jasnowski, M., and Kowalski, W. 1978. Das Eindringen von Sphagnum in kalziphile Pflanzengesellschaften in Naturschutzgebiet Tchorzyno. *Phytocoenosis* 7:71–89.

Johnstone, I. M. 1986. Plant invasion windows: A time-based classification of invasion potential. *Biological Review* 61:369–394.

Joosten, J. H. J. 1993. Denken wie ein Hochmoor: Hydrologische Selbstregulation von Hochmooren und deren Bedeutung für Wiedervernässung und Restauration. *Telma* 23:95–115.

Kirmer, A., and Mahn, E.-G. 2001. Spontaneous and initiated succession on unvegetated slopes in the abandoned lignite-mining area of Goitsche, Germany. *Applied Vegetation Science* 4:19–27.

Klimeš, L., and Klimešová, J. 2002. The effects of mowing and fertilization on carbohydrate reserves and regrowth of grasses: do they promote plant coexistence in species-rich meadows? *Evolutionary Ecology* 15:363–382.

Klötzli, F. 1987. Disturbance in transplanted grasslands and wetlands. In: *Disturbance in Grasslands*. J. van Andel, J. P. Bakker, and R. W. Snaydon (eds.). Dordrecht, Netherlands: Junk Publishers, pp. 79–96.

Koerselman, W., and Verhoeven, J. T. A. 1995. Eutrophication of fen systems: external and internal nutrient sources and restoration strategies. In: *Restoration of Temperate Wetlands*. B. D. Wheeler, S. C. Shaw, W. J. Fojt, and R. A. Robertson (eds.). Chichester: Wiley, pp. 91–112.

Lammerts, E. J., and Grootjans, A. P. 1998. Key environmental variables determining the occurrence and life span of basiphilous dune slack vegetation. *Acta Botanica Neerlandica* 47:369–392.

Lonsdale, W. M. 1999. Global patterns of plant invasions and the concept of invasibility. *Ecology* 80:1522–1536.

Lucassen, E. C. H. E. T., Smolders, A. J. P., van de Crommenacker, J., and Roelofs, J. G. M. 2004. Effects of stagnating sulphate-rich groundwater on the mobility of phosphate in freshwater wetlands: A field experiment. *Archiv für Hydrobiologie* 160:117–131.

Luken, J. O. 1990. *Directing Ecological Succession*. London: Chapman and Hall.

Marrs, R. H. 1988. Vegetation change on lowland heaths and its relevance for conservation management. *Journal of Environmental Management* 26:127–149.

Marrs, R. H. 1993. Soil fertility and nature conservation in Europe: Theoretical considerations and practical management solutions. *Advances in Ecological Research* 24:241–300.

Marrs, R. H. 2002. Manipulating the chemical environment of the soil. In: *Handbook of Ecological Restoration*. Vol. 1. M. R. Perrow and A. J. Davy (eds.). Cambridge: Cambridge University Press, pp. 155–183.

Marrs, R. H., and Bradshaw, A. D. 1993. Primary succession on man-made wastes: The importance of resource acquisition. In: *Primary Succession on Land*. J. Miles and D. W. H. Walton (eds.). Oxford: Blackwell, pp. 221–248.

Marrs, R. H., and Le Duc, M. G. 2000. Factors controlling vegetation change in long-term experiments designed to restore heathland in Breckland, UK. *Applied Vegetation Science* 3:135–146.

Marrs, R. H., Johnson S. W., and Le Duc, M. G. 1998a. Control of bracken and restoration of heathland. VIII. The regeneration of the heathland community after 18 years of continued bracken control or six years of control followed by recovery. *Journal of Applied Ecology* 35:857–870.

Marrs, R. H., Snow, C. S. R., Owen, K. M. and Evans, C. E. 1998b. Heathland and acid grassland creation on arable soils at Minsmere: Identification of potential problems and a test of cropping to impoverish soils. *Biological Conservation* 85:69–82.

Matějková, I., van Diggelen, R., and Prach, K. 2003. An attempt to restore a central European mountain grassland through grazing. *Applied Vegetation Science* 6:161–168.

McKay, J. K., Christian, C. E., Harrison, S., and Rice, K. J. 2005. "How local is local?" – A review of practical and conceptual issues in the genetics of restoration. *Restoration Ecology* 13:432–440.

Milligan, A. L., Putwain, P. D., Cox, E. S., Ghorbani, J., Le Duc, M. G., and Marrs, R. H. 2004. Developing an integrated land management strategy for the restoration of moorland vegetation on *Molinia caerulea*-dominated vegetation for conservation purposes in upland Britain. *Biological Conservation* 119:371–387.

Mitchell, R. J., Auld, M. H. D., Le Duc, M. G., and Marrs, R. H. 2000. Ecosystem stability and resilience: A review of their relevance for the conservation management of lowland heaths. *Perspectives on Plant Ecology, Evolution and Systematics* 3:142–160.

Money, R. P. 1995. Restoration of cut-over peatlands: The role of hydrology in determining vegetation quality. In: *Hydrology and Hydrochemistry of British Wetlands*. J. Hughes and L. Heathwaite (eds.). Chichester: Wiley, pp. 383–400.

Montalvo, A. M., McMillan, P. A., Allen, E. B. 2002. The relative importance of seeding method, soil ripping, and soil variables on seeding success. *Restoration Ecology* 101:52–67.

Morecroft, M. D., Masters, G. J., Brown, V. K., Clarke, I. P., Taylor, M. E., and Whitehouse, A. T. 2004. Changing precipitation patterns alter plant community dynamics and succession in ex-arable grassland. *Functional Ecology* 18:648–655.

Mouissie, A. M., Vos, P., Verhagen, H. M. C., and Bakker, J. P. 2006. Endozoochory by free-ranging, large herbivores: Ecological correlates and perspectives for restoration. Basic and Applied Ecology 6:547–558.

Munshower, F. F. 1994. *Practical Handbook of Disturbed Land Revegetation*. Boca Raton, FL: Lewis.

Myers, J. H., and Bazely, D. R. 2003. *Ecology and Control of Introduced Plants*. Cambridge: Cambridge University Press.

Ninot, J. M., Herrero, P., Ferré, A., and Guardia, R. 2001. Effects of reclamation measures on plant colonization on lignite waste in the eastern Pyrenees, Spain. *Applied Vegetation Science* 4:29–34.

Noble, I. R. 1987. The role of expert systems in vegetation science. *Vegetatio* 69:115–121.

Novák, J. and Prach, K. 2003. Vegetation succession in basalt quarries: pattern over a landscape scale. *Applied Vegetation Science* 6:111–116.

Olde Venterink, H., Wassen, M. J., Verkroost, A. W. M., and de Ruiter, P. 2003. Species richness-productivity patterns differ between N-, P-, and K-limited wetlands. *Ecology* 84:2191–2199.

Olson, D. M., and Dinerstein, E. 1998. The global 200: A representation approach to conserving the earth's most biologically valuable ecosystems. *Conservation Biology* 12:502–515.

Olsson, E. G. 1987. Effects of dispersal mechanisms on the initial pattern of old-field forest succession. *Acta Oecologica* 8:379–390.

Oomes, M. J. M., and Mooi, H. 1981. The effect of cutting and fertilizing on the floristic composition and production of an Arrhenaterion grassland. *Vegetatio* 47:233–239.

Osbornová, J., Kovářová, M., Lepš, J., and Prach, K. (eds.). 1990. *Succession in Abandoned Fields: Studies in Central Bohemia, Czechoslovakia.* Dordrecht: Kluwer.

Owen, K. M., and Marrs, R. H. 2000a. Creation of heathland on former arable land at Minsmere, Suffolk: The effect of soil acidification on the establishment of *Calluna* and ruderal species. *Biological Conservation* 93:9–18.

Owen, K. M., and Marrs, R. H. 2000b. Acidifying arable soils for the restoration of acid grasslands. *Applied Vegetation Science* 3:105–116.

Owen, K. M., Marrs, R. H., Snow, C. S. R., and Evans, C. 1999. Soil acidification—the use of sulphur and acidic litters to acidify arable soils for the recreation of heathland and acidic grassland at Minsmere, UK. *Biological Conservation* 87:105–122.

Parker, V. T. 1997. The scale of successional models and restoration objectives. *Restoration Ecology* 5:301–306.

Perkins, D. F. 1978. The distribution and transfer of energy and nutrients in the *Agrostis-Festuca* grassland ecosystem. In: *Production Ecology of British Moors and Montane Grasslands.* O. W. Heal and D. F. Perkins (eds.). Berlin: Springer, pp. 375–396.

Perrow, M. R., and Davy, A. J. (eds.). 2002. *Handbook of Ecological Restoration.* Cambridge: Cambridge University Press.

Pfadenhauer, J., Höper, H., Borkowski, B., Roth, S., Seeger, S., and Wagner, C. 2001. Entwicklung planzenartreichen Niedermoorgrünlands. In: *Ökosystemmanagement für Niedermoore; Strategien und Verfahren zur Renaturierung.* R. Kratz and J. Pfadenhauer (eds.). Stuttgart, Hohenheim: Ulmer Verlag, pp. 134–155.

Pickett, S. T. A., Collins, S. L., and Armesto, J. J. 1987. Models, mechanisms, and pathways of succession. *Botanical Review* 53:335–371.

Pickett, S. T. A., Cadenasso, M., and Bartla, S. 2001. Implications from the Buell-Small Succession Study for vegetation restoration. *Applied Vegetation Science* 4:41–52.

Poschlod, P., and Bonn, S. 1998. Changing dispersal processes in the central European landscape since the last ice age: An explanation for the actual decrease of plant species richness in different habitats? *Acta Botanica Neerlandica* 47:27–44.

Poulin, M., Rochefort, L., and Desrochers, A. 1999. Conservation of bog plant species assemblages, assessing the role of natural remnants in mined sites. *Applied Vegetation Science* 2:169–180.

Prach, K. 1985. Succession of vegetation in abandoned fields in Finland. *Annales Botanici Fennici* 22:307–314.

Prach, K. 1987. Succession of vegetation on dumps from strip coal mining, N.W.Bohemia, Czechoslovakia. *Folia Geobotanica et Phytotaxonomica* 22:339–354.

Prach, K. 2003. Spontaneous vegetation succession in central European man-made habitats: What information can be used in restoration practice? *Applied Vegetation Science* 6:125–129.

Prach, K., and Pyšek, P. 2001. Using spontaneous succession for restoration of human-disturbed habitats: Experience from central Europe. *Ecological Engineering* 17:55–62.

Prach, K., Pyšek, P., and Šmilauer, P. 1999. Prediction of vegetation succession in human-disturbed habitats using an expert system. *Restoration Ecology* 7:15–23.

Prach, K., Bartha, S., Joyce, C. B., Pyšek, P., van Diggelen, R., and Wiegleb, G. 2001. The role of spontaneous succession in ecological restoration: A perspective. *Applied Vegetation Science* 4:111–114.

Prach, K., Lepš, J., and Rejmánek, M. in press. Old field succession in Central Europe: Local and regional patterns. In: *Old fields: Dynamics and Restoration of Abandoned Farmland.* V.A. Cramer and R.J. Hobbs (eds.). Washington: Island Press.

Pyšek, P., Kučera, T., Puntieri, J., and Mandák, B. 1995. Regeneration in Heracleum mantegazzianum: Response to removal of vegetative and generative parts. *Preslia* 67:161–171.

Pyšek, P., Davis, M. A., Daehler, C. C., and Thompson, K. 2004a. Plant invasions and vegetation succession: Closing the gap. *Bulletin of the Ecological Society of America* 85:105–109.

Pyšek, P., Richardson, D. M., Rejmánek, M., Webster, G., Williamson, M., and Kirschner, J. 2004b. Alien plants in checklists and floras: Towards better communication between taxonomists and ecologists. *Taxon* 53:131–143.

Pywell, R. F., Webb, N. R. and Putwain, P. D. 1994. Soil fertility and its implications for the restoration of heathland on farmland in southern Britain. *Biological Conservation* 70:169—181.

Pywell, R. F., Webb, N. R., and Putwain, P. D. 1997. The decline of heathland seed populations following conversion to agriculture. *Journal of Applied Ecology* 34:949–960.

Pywell, R. F., Bullock, J. M., Hopkins, A., Walker, K. J., Sparks, T. H., Burke, M. J. W., and Peel, S. 2002a. Restoring of species-rich grassland on arable land: assessing the limiting processes using a multi-site experiment. *Journal of Applied Ecology* 39:249–264.

Pywell, R. F., Pakeman, R. J., Allchin, E. A., Bourn, N. A. D., Warman, E. A., and Walker, K. J. 2002b. The potential for lowland heath regeneration following plantation removal. *Biological Conservation* 108:247–258.

Pywell, R. F., Bullock, J. M., Roy, D. B., Warman, L., Walker, K. J., and Rothery, P. 2003. Plant traits as predictors of performance in ecological restoration. *Journal of Applied Ecology* 40:65–77.

Řehounková, K., and Prach, K. 2006. Spontaneous vegetation succession in disused gravel-sand pits: Role of site conditions and landscape context. *Journal of Vegetation Science.*

Rejmánek, M. 1989. Invasibility of plant communities. In: *Biological Invasions. A Global Perspective.* J. A. Drake, H. A. Mooney, F. di Castri, R. H. Groves, F. J. Kruger, M. Rejmánek, and M. Williamson (eds.). Chichester: Wiley, pp. 369–388.

Rejmánek, M. 1990. Foreword. Old and new fields of old-field ecology. In: *Succession in Abandoned Fields. Studies in Central Bohemia, Czechoslovakia.* J. Osbornová, M. Kovářová, J. Lepš, and K. Prach (eds.). Dordrecht: Kluver, pp. 9–13.

Rejmánek, M., and van Katwyk, K. P. 2004. Old-field succession: A bibliographic review (1901–1991). [http://botanika.bf.jcu.cz/suspa/]

Richardson, D. M., Pyšek, P., Rejmánek, M., Barbour, M. G., Panetta, F. D., and West, C. J. 2000. Naturalization and invasion of alien plants: Concepts and definitions. *Diversity and Distributions* 6:93–107.

Richardson, D. M., Holmes, P. M., Esler, K. J., Galatowitsch, S. M., Kirkman, S., Stromberg, J. C., Hobbs, R. J., and Pyšek, P. in press. Riparian zone: Degradation, alien plant invasions and restoration prospects. *Diversity and Distributions.*

Roberts, R. D., Marrs, R. H., Skeffington, R. A., and Bradshaw, A. D. 1981. Ecosystem development on naturally-colonized china clay wastes. I. Vegetation change and overall accumulation of organic matter and nutrients. *Journal of Ecology* 69:153–161.

Ruprecht, E. 2005. Secondary succession in old-fields in the Transylvanian Lowland (Romania). *Preslia* 77:145–157.

Ruprecht E. 2006. Successfully recovered grassland: A promising example from Romanian old-fields. *Restoration Ecology* 14:473–480.

Salonen, V., Penttinen, A., and Sarkka, A. 1992. Plant colonization of a bare peat surface, population changes and spatial patterns. *Journal of Vegetation Science* 3:113–118.

Scheffer, M. 1990. Multiplicity of stable states in fresh-water systems. *Hydrobiologia* 200:475–486.

Schmidt, W. 1983. Experimentelle Syndynamic—Neuere Wege zu einer exakten Sukzessionsforschung, dargestellt am Beispiel der Gehölzentwicklung auf Ackerbrachen. *Berichte der Deutschen Botanischen Gesellschaft* 96: 511–533.

Schmidt W. 1988. An experimental study of old-field succession in relation to different environmental factors. *Vegetatio* 77: 103–114.

Šeffer, J., and Stanová V. (eds.). 1999. *Morava River Floodplain Meadows—Importance, Restoration and Management*. Bratislava, Slovakia: Daphne.

Settele, J., Margules, C., Poschlod, P., and Henle, K. (eds.). 1996. *Species Survival in Fragmented Landscapes*. Dordrecht: Kluwer.

Shear, T. H., Lent, T. J., and Fraver, S. 1996. Comparison of restored and mature bottomland hardwood forests of southwestern Kentucky. *Restoration Ecology* 4:111–123.

Strykstra, R. J., Verweij, G. L., and Bakker, J. P. 1997. Seed dispersal by mowing machinery in a Dutch brook valley system. *Acta Botanica Neerlandica* 46:387–401.

Symonides E. 1985. Floristic richness, diversity, dominance and species evenness in old-field successional ecosystems. *Ekologia Polska* 33:61–79.

Temperton, V. M., Hobbs, R. J., Nuttle, T., and Halle, S. (eds.). 2004. *Assembly Rules and Restoration Ecology*. Washington: Island Press.

Tilman, D. 1988. *Dynamics and Structure of Plant Communities*. Princeton: Princeton University Press.

Tong, C., Le Duc, M. G., Ghorbani, J., and Marrs, R. H. 2006. Linking restoration to the wider landscape: A study of a bracken control experiment within an upland moorland landscape mosaic in the Peak District, UK. *Landscape and Urban Planning* 78(1-2):115–134.

van Andel, J., and Aronson, J. (eds.). 2006. *Restoration Ecology*. Oxford: Blackwell.

van der Hoek, D. 2005. The effectiveness of restoration measures in species-rich fen meadows. PhD-thesis, Wageningen University.

van Diggelen, R., and Grootjans, A. 1999. Restoration prospects of degraded lowland brook valleys in The Netherlands: An example from Gorecht Area. In: *An International Perspective on Wetland Rehabilitation*. W. Streever (ed.). Dordrecht: Kluwer, pp. 189–196.

van Diggelen, R., Molenaar, W. J., and Kooijman, A. M. 1996. Vegetation succession in a floating mire in relation to management and hydrology. *Journal of Vegetation Science* 7:809–820.

van Diggelen, R., Sijtsma, F. J., Strijker, D., and van den Burg, J. 2005. Relating land-use intensity and biodiversity at the regional scale. *Basic and Applied Ecology* 6:145–159.

van Duren, I. C., Pegtel, D. M., Aerts, B. A., and Inberg, J. A. 1997. Nutrient supply in undrained and drained Calthion meadows. *Journal of Vegetation Science* 8:829–838.

van Wirdum, G. 1995. The regeneration of fens in abandoned peat pits below sea level in The Netherlands. In: *Restoration of Temperate Wetlands*. B. D. Wheeler, S. C. Shaw, W. J. Fojt, and R. A. Robertson (eds.). Chichester: Wiley, pp. 251–272.

Vécrin, M. P., and Muller, S. 2003. Top-soil translocation as a technique in the re-creation of species-rich meadows. *Applied Vegetation Science* 6:271–278.

Vécrin, M. P., van Diggelen, R., Grevilliot, F., and Muller, S. 2002. Restoration of species-rich flood-plain meadows from abandoned arable fields in NE France. *Applied Vegetation Science* 5:263–270.

Verhagen, R., Klooker, J., Bakker, J. P., and van Diggelen, R. 2001. Restoration success of low-production plant communities on former agricultural soils after top-soil removal. *Applied Vegetation Science* 4:75–82.

Wali, M. K. (ed.). 1992. *Ecosystem Rehabilitation. Vol. 1: Policy Issues; Vol. 2: Ecosystem Analysis and Synthesis.* The Hague: SPB.

Walker, J., and Reuter, D. J. 1996. *Indicators of Catchment Health: A Technical Perspective.* Collingwood: CSIRO Publishing.

Walker, L. R. (ed.). 1999. *Ecosystems of Disturbed Ground. Ecosystems of the World 16.* Amsterdam: Elsevier.

Walker, L. R., and del Moral, R. 2003. *Primary Succession and Ecosystem Rehabilitation.* Cambridge: Cambridge University Press.

WallisDeVries, M. F., Bakker, J. P., and van Wieren, S. E. 1998. *Grazing and Conservation Management.* Dordrecht: Kluwer.

Wassen, M. J., van Diggelen, R., Wolejko, L., and Verhoeven, J. T. H. 1996. A comparison of fens in natural and artificial landscapes. *Vegetatio* 126:5–26.

Webb, N. R. 2002. Atlantic heathlands. In: *Handbook of Ecological Restoration.* Vol. 2. M. R. Perrow and A. J. Davy (eds.). Cambridge: Cambridge University Press, pp. 401–418.

Werger, M. J. A., Prentice, I. C., and Helspen, H. P. A. 1985. The effect of sod-cutting to different depths on *Calluna* heathland regeneration. *Journal of Environmental Management* 20:181–189.

Whisenant, S. G. 1999. *Repairing Damaged Wildlands.* Cambridge: Cambridge University Press.

Whisenant, S. G. 2002. Terrestrial systems. In: *Handbook of Ecological Restoration.* Vol. 1. M. R. Perrow and A. J. Davy (eds.). Cambridge: Cambridge University Press, pp. 83–106.

Wiegleb, G., and Felinks, B. 2001. Predictability of early stages of primary succession in post-mining landscapes of Lower Lusatia, Germany. *Applied Vegetation Science* 4:5–18.

Wolejko, L., Aggenbach, C., van Diggelen, R., and Grootjans, A. P. 1994. Vegetation and hydrology in a spring mire complex in western Pomerania, Poland. *Proceedings van de Koninklijke Nederlandse Academie van Wetenschappen* 97:219–245.

Wolf, G. 1985. Primäre Sukzession auf kiesig-sandigen Rohböden im Rheinischen Braunkohlenrevier. *Schriftenreihe für Vegetationskunde* 16:1–208.

Wolters, M., Garbutt, A., and Bakker, J. P. 2005. Recreation of intertidal habitat: Evaluating the success of de-embankments in north-west Europe. *Biological Conservation* 123:249–268.

Zavaleta, E., Hobbs, R. J., and Mooney, H. A. 2001. Viewing invasive species removal in a whole ecosystem context. *Trends in Ecology and Evolution* 16:454–459.

Zobel, M., van der Maarel, E., and Dupré, C. 1998. Species pool: The concept, its determination and significance for community restoration. *Applied Vegetation Science* 1:55–66.

7

Restoration as a Process of Assembly and Succession Mediated by Disturbance

Richard J. Hobbs, Anke Jentsch, and Vicky M. Temperton

Key Points

1. Concepts from the areas of succession, ecosystem assembly, and disturbance ecology are all interrelated and their combination can feed into a range of ideas that have relevance to ecological restoration.
2. There is a great degree of complexity to be considered in any restoration activity because that activity takes place in the context of a dynamic ecosystem with many factors affecting it, each working on a range of temporal and spatial scales. This necessitates thinking about multiple possible trajectories and restoration goals.
3. Disturbance is an important part of succession and is ongoing. While often part of the cause of the initial degradation triggering the requirement for restoration, various types of disturbance and management of the disturbance regime can also be seen as important tools in restoration.

7.1 Introduction

Successional processes in ecosystems have long been studied in ecology, and over a century of work in this field have spawned a series of different successional theories related to how ecosystems develop over time (see Chapter 1). Although ecologists agree on some of the main drivers of changes in species composition within a community, the plethora of different habitats which occur in nature, often with differing histories and organismal composition and structure, does not allow for a unifying theory of succession applicable to all ecosystems or habitats (McIntosh 1999).

When restoring degraded ecosystems we can have a range of different goals, ranging from simple reclamation of the land (e.g., reinstating any vegetation cover), to rehabilitation (reinstating some kind of ecosystem functioning such as nutrient and water cycling or productivity), to the most ambitious which is often called true restoration, where restoration of both the structure and function of the pre-disturbance state is attempted (Harris and van Diggelen 2006). Restoration goals can even include scenarios other than the pre-disturbance state accounting for novel environmental conditions and future ecosystem trajectories. In addition to concepts of ecological succession, there are other ecological theories and concepts including assembly and disturbance, which may be particularly useful when applied to degraded ecosystems that need restoring. The idea of community or ecosystem assembly arises from the observation that only certain species are able to establish and survive in any given area and that species tend to occur in recognizable and repeatable combinations or temporal sequences. Hence, there is the possibility that a set of rules can be identified that describe the processes underlying these observations. Temperton *et al.* (2004) recently evaluated the broad field of community assembly in relation to restoration. In particular, the relative merit of different approaches to assembly were assessed in relation to their direct application in restoration ecology and practice (Temperton and Hobbs 2004).

We also consider the idea that modifying ecosystem dynamics for restoration purposes can be done effectively by using disturbance as a management tool, for instance to set back the successional clock or alter abiotic restrictions for species establishment and community assembly. Additionally, natural disturbance events can be used as a "window of opportunity" for restoration purposes, such as enhancing plant establishment after heavy rainfalls associated with El Niño in arid environments (Holmgren *et al.* 2006).

Together, concepts from succession, assembly, and disturbance deal with the processes by which the living components of an ecosystem change over time and how the species assemblage present at any one time may be explained. Our premise in this chapter is that these fields, although often treated as separate entities, are complementary. Both the similarities and differences between assembly and succession mediated by disturbance can be effectively assessed to derive the most promising aspects of the fields for application in restoration. In this chapter, we thus aim to revisit perspectives from earlier chapters (particularly Chapters 5 and 6) to consider succession together with assembly and disturbance, and examine how these mesh together in the context of restoration. The aim is (1) to discuss the different types of dynamics that potentially occur in ecosystems—restored or natural, (2) to explore drivers of these dynamics, including assembly and succession modified by disturbance, and (3) to illustrate the relevance of these concepts to restoration management.

7.2   Ecosystem Dynamics

A broad categorization of the different types of dynamics that have been hypothesized to occur in ecosystems includes: deterministic dynamics, stochastic dynamics, and transitions among alternative stable stables (multiple equilibria). These are illustrated in Fig. 7.1. Essentially, an assumption of deterministic dynamics (Fig. 7.1a) suggests a predictable path of development given any particular starting point, regardless of conditions prior to or during the development of the ecosystem. This characterizes the traditional Clementsian perspective (Clements 1916), as later characterized in ecosystem terms by Odum (1969). Stochastic dynamics (Fig. 7.1b) suggest a less orderly development with the path of development constantly being affected by a variety of factors each working on a range of different temporal and spatial scales, resulting in an apparent random path. Finally, a state and transition approach (Fig. 7.1c) suggests that development is phasic and characterized by relatively sudden changes

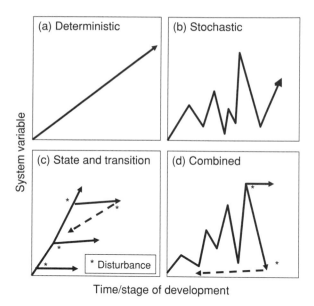

Figure 7.1 Ecosystem dynamics can be depicted as (a) deterministic, (b) stochastic, or (c) state and transition. In reality a combination of all three is likely (d).

from one stable state to another, with these changes being driven by particular events, such as various types of disturbance. Disturbance events of extreme magnitude can cause deterministic and stochastic dynamics to be nonlinear in the sense that no return to prior reference conditions or reference dynamics is possible.

The age-old debate has been whether ecosystem development is stochastic or deterministic or something in between these two. Temperton and Hobbs (2004) came to the conclusion that most ecologists would agree that simple deterministic models are just that—too simple to describe the complexity of dynamics actually observed in most situations. While some apparently deterministic patterns are observable, there are usually variations around these and often multiple developmental pathways are possible (Jentsch and Beyschlag 2003, Suding *et al.* 2004, Cramer in press, Hobbs and Walker in press). These variations often relate to the timing and severity of particular disturbances, climatic events, or soil- and resource-related phenomena.

It is increasingly clear that no one model of ecosystem dynamics is completely appropriate in all situations and for all systems. As indicated in Fig. 7.1(d), the same system may exhibit all three types of dynamics at different times, at different phases of development, or under different conditions. A further emerging realization is that whether we find stochastic, deterministic, or intermediate dynamics depends on the level of focus and on the temporal frame of reference. Fukami *et al.* (2005), for example, found community assembly rules over time when looking at plant traits but not plant species in an old field sowing experiment. Nevertheless, the recognition that different dynamic types are possible is an important one in the context of the discussion of the linkage among the different ideas surrounding succession, assembly, disturbance, and restoration. We first discuss recent developments in ideas concerning ecosystem assembly, and then return to the ideas of succession and disturbance.

7.3 Assembly

Concepts of succession and community assembly both address temporal dynamics within ecosystems. However, while succession focuses on the dynamics of a system following initial colonization of a denuded site (primary succession) or the dynamics of system regeneration after a disturbance (secondary succession), community assembly asks the question "How does the suite of species present at any particular location arrive and persist there, and how does that relate to the pool of species available within the region as a whole?" Hence, questions concerning community assembly, although inherently containing a dynamic component, are often spatially framed. Indeed, one of the main methodological approaches to assembly, the categorical approach, aims to study extant communities and provide a snapshot in time of the species, functional groups, or guilds present there. Different patterns of abundance of species in different functional groups that are found in a community are then explained via so-called assembly rules, which are often tested against null models of no interaction between organisms. An extension of this is the idea of guild proportionality, which suggests that, within a particular community type, the proportions of species of different guilds are almost constant across sites in different developmental stages (Wilson and Roxburgh 1994, Wilson 1999). As a consequence of the theory of guild proportionality, one would expect the nearest plant of a different species to belong to another guild or functional group.

Applying such concepts of community assembly to restoration situations could prove difficult, unless one could unequivocally show, for a given system, that nonconstant proportions of species in different guilds was a sign of a highly degraded site compared to a reference ecosystem exhibiting clear guild proportionality. In this case, a lack of guild proportionality in a degraded site could be used as an indication of the system being stuck in a certain state, and appropriate management measures (usually involving some kind of disturbance favoring a particular species) could be taken to move the system from the undesired stable state to a desired stable state. Although we do not have enough data on guild proportionality in different ecosystems (mainly in grasslands and deserts so far) to be able to apply such methods at this stage, it could be a promising venue for future research linking ecological theory with restoration.

A promising community assembly approach for more direct application to ecological restoration is the concept of "filters," whereby a species can only establish in an area if it can deal with the environmental conditions (i.e., the abiotic filters) as well as the other organisms it finds there (i.e., the biotic filter) (Kelt et al. 1995, Weiher and Keddy 1995, Zobel 1997, Díaz et al. 1999, but see Belyea 2004 for caveats). Various conceptualizations of environmental "filters," for example very low or high nutrient levels in an ecosystem, also tend to indicate that the main effect of filters is to vary the species composition in relation to environmental (and hence spatial) variation (Díaz et al. 1999; Hobbs 2004). On the other hand, the "response" or dynamic approach to assembly considers changes in the biotic community and the expression of community assembly "rules" over time, and recent treatments emphasize the dynamic nature of filters, which are likely to change over time as well as spatially (Fattorini and Halle 2004; Hobbs 2004). As discussed in Temperton and Hobbs (2004), the dynamic filter approach could prove useful, at least before the onset of restoration measures, to help assess what the limiting factors for restoration are, and how one could alter the filter mesh to allow certain species to establish.

This is where the issue of disturbance comes in. Disturbances modify community assembly in two ways. Firstly, disturbances change environmental filters such as nutrient availability, and secondly, they act on plant traits as filters of species (biotic) assembly in their own right (White and Jentsch 2004). Thus, disturbance can be not only the cause of degradation, but at appropriate scales and magnitude can also be a direct tool for restoration managers wanting to restore appropriate ecosystem dynamics. An example here would be the restoration of species-rich calcareous grasslands on nutrient-poor sites, where active disturbance in the form of sheep grazing or mowing of the grassland, keeps the ecosystem from accumulating nutrients and from undergoing succession to shrub- and woodland. Another example would be the restoration of species-poor, resource-limited inland sand dunes, where, at least in some situations, anthropogenic disturbances in the form of military maneuvers or top-soil removal keep parts of the ground bare for seedling establishment and again prevent the ecosystem from accumulating nutrients and undergoing succession to shrub- and woodland (Jentsch and Beyschlag 2003).

7.4 Succession—Not One-Way

Although there are various conceptual models in existence, succession is widely acknowledged to be a continuous, though often stepwise, process of species turnover with varying speeds and trajectories (see Chapter 1). Thus, the application of knowledge about temporal dynamics in ecosystems is the fundamental approach in successional theory that can be linked to restoration. We explore the idea that temporal dynamics in ecosystems are the product of two interacting factors: continuous versus discrete processes (Jentsch and White, unpublished data). Continuous processes include gradual accumulation of biomass and nutrients, as the system moves through progressive successional stages. Discrete processes include the occurrence of disturbance, which can cause rapid transitions between different ecosystem states or suddenly reset the successional clock. In addition, disturbance can change continuous processes such as colonization or extinction of indigenous species to sudden events such as rapid invasion of an alien species. Thus, restoration managers have two different options for modifying ecosystem dynamics at restoration sites: manipulating continuous processes (succession) or making use of discrete events (disturbance).

While most successional sequences seem to follow a progressive accumulation of biomass, successional studies on very long-term chronosequences or pollen sequences indicate a decline phase in succession in the absence of a major disturbance (Iversen 1969, Wardle *et al.* 2004). This mirrors earlier considerations of vegetation dynamics as more cyclical than unidirectional (Watt 1947, Remmert 1991, van der Maarel 1996), and is included in more recent conceptualizations of ecosystem dynamics (Holling 1986 and 1993). These ideas of cyclic dynamics implicitly include a decline phase which is followed by a recovery phase, either following disturbance or after the slow release of accumulated biomass and nutrients.

The concept of non unidirectional succession allows a reconsideration of theories of succession as initially formulated at the beginning of the 20th century, in particular the dichotomy between deterministic climax models (Clements 1916) and stochastic individualistic models of terrestrial succession (Gleason 1926). In part, the recognition of a decline phase in succession after reaching a phase of maximum biomass accumulation or species richness could be viewed as providing support for Clements' view that ecosystems can develop quite deterministically over time toward a convergent state with a maximum biomass or a stable end point. However, retrogression can happen in different ways and at different times in various seres. Perhaps two distinct categories of retrogressive successions can be recognized: "natural" retrogressions, such as that observed on the Coooloola dunes in Australia (see Chapter 4), and human-induced retrogressions arising from altered grazing and fire regimes, mining, and the like (see Chapters 2, 4, and 6)

On the other hand, this pattern and the subsequent decline phase can result from the individualistic growth and maturation of particular dominant species, indicating that Clements' emergent properties can be explained using Gleason's individualistic approach. Intriguingly, recent work on the assembly of vegetation on old fields over time (Fukami *et al.* 2005) found convergence of functional plant traits over time—indicating more deterministic dynamics—but divergence of species composition as a result of priority effects—indicating more stochastic dynamics. This work raises important issues of how the level of focus of our research affects the outcome, i.e., how dependent is our hypothesis-testing of different ecological theories on community succession and assembly on the unit of focus, be it species, community structure, functional groups, or species traits? To ensure more cross-fertilization between community ecology and restoration, in the future we need to make sure we try to answer the critical questions at several levels of focus. Otherwise, we may be unwittingly missing out on crucial information for restoring ecosystems. It could well be that continued debate on deterministic *versus* stochastic theories of ecosystem is in part due to the possibility that both theories (as well as the theory of alternative stable states) can apply at the same time (depending on the level of focus) or within the same system at different times. Resolving this major issue in ecology, by knowing at what level of focus one or another theory applies, would help greatly improve the applicability of ecological theory to restoration.

In succession, the lack of extensive examination of the decline phase after the maximum biomass stage is probably due to the shorter timescale of earlier studies. Most studies have been carried out on relatively young landscapes, lack data from very long-term chronosequences, and illustrate a bias of successional focus on build-up and not break-down concepts (see also Chapter 4). Experience in restoration projects has repeatedly shown, however, that the story of successional direction is complex, particularly in a world experiencing drastic changes in disturbance regimes (Hölzel and Otte 2003, Jentsch and Beierkuhnlein 2003, Aronson and Vallejo 2006). Disturbance and its interactive effects on species colonization and extinction potential are a vital part of the dynamics of ecosystem development. Disturbance regimes, including intensity and frequency of disturbance events, are recognized as critical drivers of successional trajectories (Trudgill 1977, White and Jentsch 2001).

We often see restoration as an acceleration of succession via human intervention after major or chronic human disturbance. The issue is whether the highly degraded systems, called in older parlance "dis-climax" systems, can be restored or naturally regenerate back to a state similar to that prior to the major disturbance. Restoration can learn from how succession proceeds on highly degraded systems. Degraded, naturally retrogressive seres can keep going downhill, as can a system subjected to an anthropogenic disturbance. So, is restoration trying to stop, reverse, and/or completely restart the succession (see Chapters 4 and 5)?

The key aspects which do not allow for simple acceleration of successional processes in restoration are generally related to threshold phenomena which occur when a system has been degraded beyond its resilience or inherent capacity to recover (Whisenant 1999, Hobbs and Harris 2001, Harris and van Diggelen 2006). Such thresholds, often difficult to identify and quantify before they are crossed, may be biotic in origin, relating to species loss, gain, or altered dispersal potential, or abiotic, relating to changes in the physical or chemical characteristics of the environment (for instance, altered soil structure or chemistry). The presence of such thresholds militates against a simple successional process and results instead in the possibility of alternative states with the system "stuck" in a particular state (at least in the time frames within which humans operate) with little or no potential for further development without active intervention to overcome the threshold phenomenon in evidence.

Another complication of accelerated succession as a simple restoration goal is ongoing, irreversible change in the environment, such as accumulating atmospheric nitrogen deposition. Such change may create novel landscapes in the sense of environmental determinants for species assembly and successional interaction. Often, restoration goals cannot rely on historical reference seres; rather, they have to account for novel environments.

7.5 Combining Succession and Assembly

The succession and community assembly approaches can be viewed as complementary and inherently have a lot in common. Clearly, the more mechanistic models of succession, such as those described by Connell and Slatyer (1977) and Noble and Slatyer (1980), can be interpreted in a community assembly framework with observed dynamics being the result of impacts of various events on the response of individual species and interactions among species. The inhibition, facilitation, and tolerance models of Connell and Slatyer (1977), for example, have a lot in common with so-called "priority effects" in the field of community assembly rules (Drake 1991, Belyea and Lancaster 1999) whereby the identity of the established species within an ecosystem has an effect on newcomers to the system. Priority effects in assembly can also be negative (competitive exclusion), positive (nurse–plant effects), or neutral (both positive and negative) depending on, for example, the phenological phase of plants interacting under nurse–plant conditions (e.g., Flores-Martínez et al. 1994). Hence, concepts such as priority effects could simply be viewed as the renaming of older concepts from the succession literature.

Similarly, the ideas of community assembly can be interpreted in a successional context, especially where community assembly is considered in a temporal context and abiotic as well as biotic factors are considered. This is increasingly the focus of attention in community assembly studies (see Weiher and Keddy 1999), even though there is disagreement among ecologists as to whether community assembly rules should include or exclude abiotic, environmental factors (cf. Temperton and Hobbs 2004). As such, dynamic models of community assembly, such as the dynamic filter model proposed by Fattorini and Halle (2004), form a very useful link between the focus of community assembly on biotic interactions within communities and the focus of succession on temporal dynamics of the biotic and abiotic components of a system after a disturbance. In summary, succession focuses on what is at a site or able to colonize while community assembly focuses on what does and does not get established, but both are concepts of change in communities over time.

Rather than continue with ecology's traditional propensity to fragment its subject matter instead of synthesizing disparate approaches, one could argue that assembly ideas differ very little from succession concepts and hence should not be considered as a separate entity. Despite the different origin of assembly ideas, this seems a profitable way to proceed, at least where assembly is considered in a dynamic context. Indeed, recent studies appear to assume that assembly and succession can be considered together to help explain observed patterns and dynamics (Bossuyt *et al.* 2005).

7.6 Disturbance

Lockwood (1997) and Young *et al.* (2001) both suggest that succession could form the core concepts in restoration. We would like to go further and suggest that disturbance, assembly, and succession together form the key to a better understanding of concepts in restoration. This may at first sight seem trivial, as we all know that the main reasons for an ecosystem needing restoration are usually caused by anthropogenic or natural disturbances, but we propose to focus particularly on the differential effects of disturbance type, frequency, and intensity on successional trajectories, on the establishment and extinction of species within a system, and on assembly rules of community dynamics under current disturbance regimes. The idea that disturbance is an important part of succession is, of course, not new (e.g., Pickett and White 1985), but we suggest that a more coherent combination of ideas from succession, assembly, and disturbance may be beneficial.

According to Pickett and White (1985), we define disturbance in a neutral way as a discrete event in time that disrupts the ecosystem, community, or population structure, and changes the resources, substrate availability, or the physical environment. Disturbance in a restoration context is far more than just the event that creates the degradation or change of state. Disturbance can be an essential tool of management during the restoration process itself, because it can modify ecosystem dynamics. Disturbances include a wide variety of events at different spatial and temporal scales, ranging from small-scale animal diggings through fires and floods to broad-scale storms and tectonic and volcanic activity. Additionally, disturbance regimes, or the mix of different disturbances characterized by their size, frequency, and intensity, need to be restored as such, because they play a crucial role in dynamics of restored sites in many ecosystems (see Chapters 4 and 6).

Although many restoration ecologists may conceive of disturbance as limited to the period before restoration begins, disturbance also helps to produce the continuing dynamics that control the community assembly—establishment and turnover—of individuals and the successional dynamics of communities. For instance, disturbances act directly as filters on plant traits (Keddy 1992, Díaz *et al.* 1999, White and Jentsch 2004), e.g., on survival, reproduction, colonization, or dispersal traits. In addition, disturbances are dynamic mechanisms that modify the process of community assembly through time, e.g., by modifying competitive interactions, successional pathways, and trajectories (Noble and Slatyer 1978 and 1980). Disturbances also modify the process of invasion by alien species (see Chapters 3 and 6), which may lead to sudden changes in successional pathways. For example, introduced grasses in western North America (Billings 1990) and invasive exotic trees in the Florida Everglades have substantially modified the fire regime in that they increased fire frequency and intensity (Bodle *et al.* 1994), thereby modifying the limiting disturbance filter for community assembly. Disturbance-mediated invasion calls for flexible management strategies (Huston 2004).

In summary, to help inform restoration projects, we need to learn more about the impacts of different kinds of disturbance as a tool in restoration: which disturbance and how much of it produces desired effects (such as native species establishment) and which disturbance tends to favor the invasion of alien species. Much of this information will, of course, depend on the surrounding species pool at local and regional levels, but inevitably also on the disturbance regime of the site.

7.7 Applying Assembly, Succession, and Disturbance Concepts to Restoration

In order to be applicable to restoration, theoretical considerations need to reach a certain degree of internal consistency, generality, and proven applicability in particular systems. There are a variety of ways in which the relationship between assembly, succession, and disturbance concepts and the practice of restoration can be depicted (Fig. 7.2). At worst, each area can be visualized as being developed and pursued in complete isolation from the others (Fig. 7.2a). Alternatively, we would argue that the concepts are increasingly being linked in meaningful ways. One interpretation of this is that succession and ecosystem assembly are nonoverlapping sets of ideas, but that the processes encompassed by each are mediated by disturbance and in this way feed into restoration practice (Fig. 7.2b). Another viewpoint is that succession and assembly share some concepts and ideas in common (Fig. 7.2c), while a more extreme view might be that assembly is simply a current "bandwagon" that merely restates some of the concepts originally developed in a successional context (Fig. 7.2d). We do not offer a definitive statement on which of these viewpoints is correct, because this partially depends on the experience and training of the observer.

What we do suggest, however, is that such discussions actually have little relevance to the practice of restoration. The practitioner really will not care too much about the finer arguments distinguishing between succession and assembly, and will instead want to know what the theoretical discussions mean for what actually has to be done in real-world applications. Therefore, we need to get our scientific house in order and clarify where concepts and theories can provide useful input to these applications. In this chapter, we have attempted

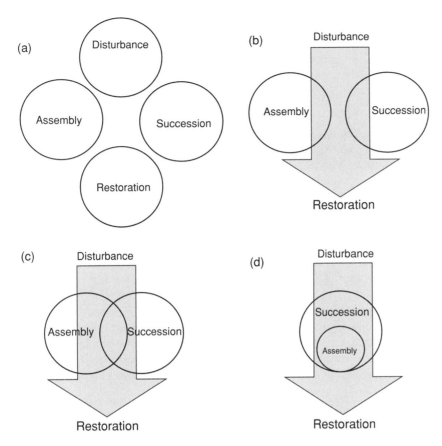

Figure 7.2 Ways of conceptualizing the links between ideas on succession, assembly, disturbance, and restoration. (a) Each is separate and does not interact with the other, (b) Assembly and succession are separate bodies of theory, but each is mediated by disturbance and has relevance to restoration, (c) As for (b) except that assembly and succession overlap in some aspects, (d) Assembly is seen as a subset of succession theory.

to indicate where profitable linkages between the three sets of concepts of succession, assembly, and disturbance may lie. Further development of these potential synergies may form part of a conceptual "toolbox" which can usefully inform restoration practice (Fig. 7.3). In this final section, we provide a series of suggestions for issues which might be important in this regard.

Managing disturbance regimes to modify community assembly and succession in a restoration process obviously poses some fundamental challenges to practitioners. Human disturbance regimes, e.g., due to management action such as mowing, grazing, or suppression of fire, need to be tested for their historic bounds of variation. Disturbances that have historic precedence and hence produce conditions that are within the historic bounds of variation for an ecosystem may produce different responses, or serve different restoration goals, than disturbances that are novel or create conditions that are outside those bounds (White and Jentsch 2004). Also, novel disturbances (ones not previously experienced by the ecosystem) act as filters for community assembly. At evolutionary time scales, precedence would ultimately be responsible for the range of life

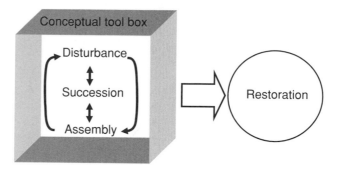

Figure 7.3 Ideally, concepts from succession, assembly, and disturbance should be combined synergistically to produce a useful body of ideas which have direct relevance to restoration practice.

histories present and the occurrence of species adapted to the disturbance—and hence which species are available for community assembly and successional dynamics.

We list a series of considerations that arise from the above discussions which relate to the practicalities of restoration. They include considerations of historical precedence, continuous versus discrete processes, self-sustainable dynamics, thresholds, restoration goals and alternative states, scaling issues, spatial mosaics, nonlinear dynamics, inertia, and underlying processes for management action.

1. *Precedence.* Historical contingency (White and Jentsch 2001) needs to be taken into account in restoration management. Only those species that have access to the site and traits to pass the disturbance filter (in addition to the traits needed to pass other biotic and abiotic filters) can participate in recovery and assembly. Thus, historical precedence and the regional species pool determine the diversity of functional responses within a restoration site.

2. *Continuous and discrete processes.* In essence, the interaction of continuous and discrete processes drives successional dynamics. Thus, the interaction of continuous and discrete processes implies gradual accumulation as well as sudden change in resources, and defines options of human intervention in restoration. Patterns of seasonality and their suddenness add to the complexity of successional rhythms in various restoration sites. These seasonal patterns define, or at least interact with, the temporal rhythm of discrete events (disturbances) and continuous processes (e.g., growth, community assembly, and regeneration). These patterns thereby define the appropriate timing for management action.

3. *Self-sustainable dynamics.* No matter where restoration projects try to apply successional theory, the primary goal of restoration is to apply knowledge about the conditions that establish self-sustaining dynamics (or autogenic processes) within a particular ecosystem. In restoration practice, this knowledge may define thresholds in the above-mentioned continuous processes, such as increasing resource availability, where competitive balance of interacting species is altered, or in discrete events, such as repeated droughts, where composition of communities is altered. Continuous processes such

as nutrient accumulation in soils, a common problem in the highly industrialized countries of northern Europe, can seriously alter the species
composition of ecosystems, despite the overall species richness remaining
very similar over time. To reach a restoration goal of reinstating certain
plant communities adapted to low nutrient conditions (such as calcareous
grasslands or heathlands), one therefore has to actively disturb the process
of continuous accumulation of nutrients, for instance by removing topsoil
(see Chapters 5 and 6). To allow the desired community to assemble, these
actions have to be carried out in close vicinity to viable populations of the
target species, or even in conjunction with introducing the desired species.

4. *Thresholds*. The restoration goal of self-sustaining dynamics suggests that
   we can find thresholds for nonsustainable dynamics which lead to changes
   in state (recently also referred to as regime shifts; Scheffer and Carpenter
   2003; Mayer and Rietkerk 2004). Within the area of management actions
   using disturbance as a tool for modifying ecosystem dynamics, thresholds leading to nonsustainable dynamics should not be crossed. Therefore,
   restoration ecologists need to understand the interplay of discrete versus
   continuous processes and to carefully manage for a dynamic balance between the two. This is, of course, no simple task, and developing an ability
   to identify, and hence avoid, thresholds of nonsustainable dynamics for different systems forms a current challenge for ecologists and restorationists
   alike.

5. *Restoration goals and alternative stable states.* Restoration projects can
   and should include multiple goals and the maintenance of a variety of successional states, and thus several reference dynamics (Grimm and Wissel
   1997) also referred to as "multiple equilibria" (Scheffer and Carpenter 2003;
   Suding *et al.* 2004). Considering that restoration projects often aim at restoring specific types of habitat (such as calcareous grassland or heathland),
   this can pose a practical problem. One way of dealing with the possibility of
   "multiple equilibria" is to include as much of a mosaic of different habitats/
   vegetation patches within a restoration site as possible, to allow for shifts in
   states from one successional stage or alternative stable state to the next (see
   Chapter 2). Alternatively, active management is usually needed to keep a
   habitat in a certain desired state. In addition to the influence of management,
   changing environmental conditions, including disturbances as well as more
   continuous environmental changes, may contribute to shifting ecosystems
   between alternative states (Suding *et al.* 2004). Within a particular state,
   resilience is mainly determined by the interplay of disturbances with internal system feedbacks, e.g., in terms of disturbance magnitude, which can
   be absorbed by the system without changing (Mayer and Rietkerk 2004).

6. *Scaling issues*. In practice, disturbance regimes are important determinants
   of both restoration site trajectories including self-sustaining dynamics and
   state shifts. For example, Turner *et al.* (1993) introduced the concept of
   landscape equilibrium caused by various kinds of disturbance regimes.
   They predict that the presence or absence of equilibrium and variance in an
   ecosystem is defined by the dimensions of disturbance relative to landscape
   extent and speed of successional dynamics. If the ratio of disturbance area
   to landscape area, or the ratio of disturbance frequency to the time needed
   for successional recovery, is very large, single, not necessarily dramatic
   disturbance events may destabilize the dynamic equilibrium of one regime

and reach the threshold for bifurcation and regime shift. As for disturbance intensity, some disturbances result in straightforward secondary succession that reestablishes the pre-disturbance composition, structure, and resources, whereas others affect site quality through long-term decreases or increases in resource levels, leading to trajectories that are out of bounds of the pre-disturbance situation (Walker and del Moral 2003). In the state of equilibrium, there is bounded variation: in a large enough restoration area, no species or successional states become extinct across the area as a whole (although they may do so in individual patches), but they can fluctuate in abundance due to the impact of disturbances. For example, species with concordant life cycles dominate periods right after a disturbance event, while species with discordant life cycles dominate later phases of succession (Pavlovic 1994). Nevertheless, both kinds of species contribute to community assembly in a disturbance-prone ecosystem. Understanding disturbance effects and subsequent system responses is crucial for understanding regime shifts and their thresholds and scales. Here, disturbance ecology offers valuable pointers to be applied in restoration action, and more cooperation between the ecologists and restorationists on this issue will no doubt provide more concrete guidelines for action in the future.

7. *Spatial mosaics.* Restoration goals are not static. A crucial challenge for restoration ecology is therefore to understand ecosystem dynamics as a function of spatial and temporal interrelations of different elements of the disturbance regime. Disturbances and management action play a crucial role for initiating and stabilizing successional rhythms (Jentsch *et al.* 2002a). If a restored site is all in one age state, regardless of whether this is recently disturbed or long undisturbed, it will lose species that characterize the other age states (Pickett and Thompson 1978; Beierkuhnlein and Jentsch 2005) and thus will also lose its ability to respond to disturbance (Jentsch 2004). Restoration managers need to plan for a sustainable mosaic of all stages and species (Pickett and Thompson 1978) and for dynamic processes sustaining this mosaic.

8. *Non-linear dynamics.* Although we can direct restoration to a certain extent via knowledge of community assembly and succession in combination with the appropriate disturbance regime, there will inevitably also be an added factor of unpredictability within an ecosystem. Internal feedbacks within ecosystems can interact with broad-scale external forces, such as global weather patterns or restoration efforts, and trigger shifts to either alternative regimes or to novel trajectories. Nonlinear system dynamics imply that a system's retransformation leads to novel conditions instead of prior structures and functions (Beisner *et al.* 2003, Holmgren *et al.* 2006). Often, nonlinearity of ecosystem dynamics or regimes shifts are neither very obvious nor dramatic. For example, factors that undermine resilience slowly, such as eutrophication in resource-limited systems (Verhagen *et al.* 2001, Jentsch *et al.* 2002b, Hölzel and Otte 2003), disturbance-mediated introduction of invasive species (Sharp and Whittaker 2003) or climate change (Jentsch and Beierkuhnlein 2003), can be responsible for altered successional trajectories.

9. *Inertia.* Although in some cases changes in state can occur suddenly, attention of restoration managers needs to be drawn to gradual accumulation up to thresholds of state change. Climate change is one of the major driving

forces that contributes to alterations in disturbance regime, e.g., change in fire frequency due to variations in weather conditions, or increased flooding intensity due to altered precipitation patterns. Such disturbances can then remove the inertia present in existing ecosystems, resulting in a relatively sudden response (or adjustment) to novel conditions.

10. *Underlying processes.* Successional pathways can be irreversibly altered in terms of composition and velocity when modified by disturbance regimes, and exposed to gradually or suddenly varying environmental conditions. Thus, another crucial challenge in restoration ecology is to assess whether restoration efforts within an ecosystem, such as manipulation of the disturbance regime, interact with underlying continuous processes. A crucial insight here is that both external triggering factors and internal variables may be slow or fast, can occur independently from each other, and can be disjointed over temporal and spatial scales (Trudgill 1977). Our challenge is to assess whether management action and underlying processes are in synchrony and harmony, or whether counter-effects will occur.

The ability to address successfully issues such as those listed above will require the use and integration of all available conceptual and practical tools. In this chapter, we have attempted to integrate the concepts from classical successional approaches and emerging approaches of community assembly. A key ingredient in doing this is the recognition of the importance of disturbance and alternative pathways and stable states in ecosystem dynamics. The overall goal of restoration can be viewed as the modification of ecosystem dynamics so that the system moves toward some stated target condition. In order to do this, we need to recognize that the dynamics can be deterministic, stochastic, or switching between alternative states; the challenge is to identify which type of dynamic is in play and to then decide what to do about it. The tools to do something about it will include manipulating the processes of community assembly (determining what gets there) and succession (what happens subsequently), and doing so by managing and using disturbance of various types. We suggest that the improved understanding of dynamics resulting from the integration of succession, assembly, and disturbance concepts can lead to the development of more realistic restoration goals and more effective restoration practice.

Acknowledgments: We thank Joe Walker, Lawrence Walker, and several anonymous referees for valuable comments on the draft manuscript. We also thank Peter White for earlier discussions on the role of disturbance in community assembly.

References

Aronson, J., and Vallejo, R. 2006. Challenges for the practice of ecological restoration. In: *Restoration Ecology: The New Frontier.* J. van Andel and J. Aronson (eds.). Oxford: Blackwell, pp.234–247.

Beierkuhnlein, C., and Jentsch, A. 2005. Ecological importance of species diversity. A review on the ecological implications of species diversity in plant communities. In: *Plant Diversity and Evolution: Genotypic and Phenotypic Variation in Higher Plants.* B. R. Henry, (ed.). Wallingford: CAB International, pp. 249–285.

Beisner, B. E., Haydon, D. T., and Cuddington, K. 2003. Alternative stable states in ecology. *Frontiers in Ecology and Environment* 1:376–382.

Belyea, L. R. 2004. Beyond ecological filters: Feedback networks in the assembly and restoration of community structure. In: *Assembly Rules and Restoration Ecology: Bridging the Gap Between Theory and Practice.* V. M. Temperton, R. J. Hobbs, T. J. Nuttle and S. Halle (eds.). Washington, D.C.: Island Press, pp. 115–131.

Belyea, L. R., and Lancaster, J. 1999. Assembly rules within a contingent ecology. *Oikos* 86:402–416.

Billings, W. D. 1990. *Bromus tectorum*, a biotic cause of ecosystem impoverishment in the Great Basin. In: *The Earth in Transition.* G. M. Woodwell (ed.). Cambridge: Cambridge University Press, pp. 301–322.

Bodle, M. J., Ferriter, A. P., and Thayer, D. D. 1994. The biology, distribution, and ecological consequences of *Melaleuca quinquenerva* in the Everglades. In: *Everglades: The Ecosystem and Its Restoration.* S. Davis and J. Ogden (eds.). St Lucia: St Lucia Press, pp. 341–355.

Bossuyt, B., Honnay, O., and Hermy, M. 2005. Evidence for community assembly constraints during succession in dune slack plant communities. *Plant Ecology* 178:201–209.

Clements, F. E. 1916. *Plant Succession: An Analysis of the Development of Vegetation.* Washington, D.C.: Carnegie Institution of Washington Publication Number 242.

Connell, J. H., and Slatyer, R. O. 1977. Mechanisms of succession in natural communities and their role in community stability and organization. *The American Naturalist* 111:1119–1144.

Cramer, V. A. in press. Old field dynamics revisited: Alternative stable states. In: *Old Fields: Dynamics and Restoration of Abandoned Farmland.* V. A. Cramer and R. J. Hobbs (eds.). Washington, D.C.: Island Press.

Díaz, S., Cabido, M., and Casanoves, F. 1999. Functional implications of trait-environment linkages in plant communities. In: *Ecological Assembly Rules: Perspectives, Advances, Retreats.* E. Weiher and P. Keddy (eds.). Cambridge: Cambridge University Press, pp. 338–362.

Drake, J. A. 1991. Community-assembly mechanics and the structure of an experimental species ensemble. *The American Naturalist* 137:1–26.

Fattorini, M., and Halle, S. 2004. The dynamic environmental filter model: How do filtering effects change in assembling communities after disturbance? In: *Assembly Rules and Restoration Ecology: Bridging the Gap Between Theory and Practice.* V. M. Temperton, R. J. Hobbs, T. J. Nuttle, and S. Halle (eds.). Washington, D.C.: Island Press, pp. 96–114.

Flores-Martinez, A., Ezcurra, E., and Sanchez-Colon, S. 1994. Effects of *Neobuxbaumia tetetzo* on growth and fecundity of its nurse plant *Mimosa luisana. Journal of Ecology* 82:325–330.

Fukami, T., Bezemer, T. M., Mortimer, S. R., and van der Putten, W. H. 2005. Species divergence and trait convergence in experimental plant community assembly. *Ecology Letters* 8:1283–1290.

Gleason, H. A. 1926. The individualistic concept of the plant association. *Bulletin of the Torrey Botanical Club* 53:7–26.

Grimm, V., and Wissel, C. 1997. Babel, or the ecological stability discussions: An inventory and analysis of terminology and a guide for avoiding confusion. *Oecologia (Berlin)* 109:323–334.

Harris, J. A., and van Diggelen, R. 2006. Ecological restoration as a project for global society. In: *Restoration Ecology: The New Frontier.* J. van Andel and J. Aronson (eds.). Oxford: Blackwell, pp. 3–15.

Hobbs, R. J. 2004. Ecological filters, thresholds and gradients in resistance to ecosystem reassembly. In: *Assembly Rules and Restoration Ecology: Bridging the Gap Between*

Theory and Practice. V. M. Temperton, R. J. Hobbs, T. J. Nuttle, and S. Halle (eds.). Washington, D.C.: Island Press, pp. 72–95.

Hobbs, R. J., and Harris, J. A. 2001. Restoration Ecology: Repairing the Earth's ecosystems in the new millennium. *Restoration Ecology* 9:239–246.

Hobbs, R. J., and Walker, L. in press. Old field succession: Development of concepts. In: *Old Fields: Dynamics and Restoration of Abandoned Farmland.* V. A. Cramer and R. J. Hobbs (eds.). Washington, D.C.: Island Press.

Holling, C. S. 1986. Resilience of ecosystems; local surprise and global change. In: *Sustainable Development of the Biosphere.* W. C. Clark and R. E. Munn (eds.). Cambridge: Cambridge University Press, pp. 292–317.

Holling, C. S. 1993. Investing in research for sustainability. *Ecological Applications* 3:552–555.

Holmgren, M., Stapp, P., Dickman, C. R., Gracia, C., Graham, S., Gutiérrez, J. R., Hice, C., Jaksic, F., Kelt, D. A., Letnic, M., Lima, M., Lopez, B. C., Meserve, P. L., Milstead, W. B., Polis, G. A., Previtali, M. A., Richter, M., Sabaté, S., and Squeo, F. A. 2006. Extreme climatic events shape arid and semiarid ecosystems. *Frontiers in Ecology and Environment* 4:87–95.

Hölzel, N, and Otte A. 2003. Restoration of species-rich flood-meadow by topsoil removal and diaspore transfer with plant material. *Applied Vegetation Science* 6:131–140.

Huston, M. 2004. Management strategies for plant invasion: Manipulating productivity, disturbance, and competition. *Diversity and Distributions* 10:167–178.

Iversen, J. 1969. Retrogressive development of a forest ecosystem demonstrated by pollen diagrams from fossil mor. *Oikos 12* (Suppl.) :35–49.

Jentsch, A. 2004. Disturbance driven vegetation dynamics. Concepts from biogeography to community ecology, and experimental evidence from dry acidic grasslands in central Europe. *Dissertationes Botanicae* 384:1–218.

Jentsch, A., and Beierkuhnlein, C. 2003. Global climate change and local disturbance regimes as interacting drivers for shifting altitudinal vegetation patterns in high mountains. *Erdkunde* 57:218–233.

Jentsch, A., and Beyschlag, W. 2003. Vegetation ecology of dry acidic grasslands in the lowland area of central Europe. *Flora* 198:3–26.

Jentsch, A., Beierkuhnlein, C., and White, P. S. 2002a. Scale, the dynamic stability of forest ecosystems and the persistence of biodiversity. *Silva Fennica* 36:393–400.

Jentsch, A., Friedrich, S., Beyschlag, W., and Nezadal, W. 2002b. Significance of ant and rabbit disturbances for seedling establishment in dry acidic grasslands dominated by *Cornyephorus canescens. Phytocoenologia* 32:553–580.

Keddy, P. A. 1992. Assembly and response rules: Two goals for predictive community ecology. *Journal of Vegetation Science* 3:157–164.

Kelt, D. A., Taper, M. L., and Meserve, P. L.1995 Assessing the impact of competition on community assembly: A case study using small mammals. *Ecology* 76:1283–1296.

Lockwood, J. L. 1997. An alternative to succession: Assembly rules offer guide to restoration efforts. *Restoration and Management Notes* 15:45–50.

Mayer, A. L., and Rietkerk, M. 2004. The dynamic regime concept for ecosystem management and restoration. *BioScience* 54:1013–1020.

McIntosh, R. P. 1999. The succession of succession: A lexical chronology. *Bulletin of the Ecological Society of America* 80:256–265.

Noble, I. R., and Slatyer, R. O. 1978. The effect of disturbance on plant succession. *Proceedings of the Ecological Society of Australia* 10:135–145.

Noble, I. R., and Slatyer, R. O. 1980. The use of vital attributes to predict successional changes in plant communities subject to recurrent disturbance. *Vegetatio* 43:5–21.

Odum, E. P. 1969. The strategy of ecosystem development. *Science* 164:262–270.

Pavlovic, N. B. 1994. Disturbance-dependent persistence of rare plants: Anthropogenic impacts and restoration implications. In: *Recovery and Restoration of Endangered*

Species. M. L. Boels and C. Whelan (eds) Cambridge: Cambridge University Press, pp. 159–193.

Pickett, S. T. A., and Thompson, J. N. 1978. Patch dynamics and the design of nature reserves. *Biological Conservation* 13:27–37.

Pickett, S. T. A., and White, P. S. 1985. Natural disturbance and patch dynamics: An introduction. In: *The Ecology of Natural Disturbance and Patch Dynamics.* S. T. A. Pickett and P. S. White (eds.). Orlando: Academic Press, pp. 3–13.

Remmert, H. (ed.). 1991. *The Mosaic Cycle Concept of Ecosystems.* New York: Springer.

Scheffer, M., and Carpenter, S. R. 2003. Catastrophic regime shifts in ecosystems: Linking theory to observation. *Trends in Ecology and Evolution* 18:648–656.

Sharp, B. R., and Whittaker, R. J. 2003. The irreversible cattle-driven transformation of a seasonally flooded Australian savanna. *Journal of Biogeography* 30:783–802.

Suding, K. N., Gross, K. L., and Houseman, G. R. 2004. Alternative states and positive feedbacks in restoration ecology. *Trends in Ecology and Evolution* 19:46–53.

Temperton, V. M., and Hobbs, R. J. 2004. The search for ecological assembly rules and its relevance to restoration ecology. In: *Assembly Rules and Restoration Ecology: Bridging the Gap Between Theory and Practice.* V. M. Temperton, R. J. Hobbs, T. J. Nuttle, and S. Halle (eds.). Washington, D.C.: Island Press, pp. 34–54.

Temperton, V. M., Hobbs, R. J., Nuttle, T. J., and Halle, S.(eds.). 2004. *Assembly Rules and Restoration Ecology: Bridging the Gap Between Theory and Practice.* Washington, D.C.: Island Press.

Trudgill, S. T. 1977. *Soil and Vegetation Systems.* Oxford: Clarendon Press.

Turner, M. G., Romme, W. H., Gardner, R. H., O'Neill, R. V., and Kratz, T. K. 1993. A revised concept of landscape equilibrium: Disturbance and stability on scaled landscapes. *Landscape Ecology* 8:213–227.

van der Maarel, E. 1996. Pattern and process in the plant community: Fifty years after A. S. Watt. *Journal of Vegetation Science* 7:19–28.

Verhagen, R., Klooker, J., Bakker, J. P., and van Diggelen R. 2001. Restoration success of low-production plant communities on former agricultural soils after top-soil removal. *Applied Vegetation Science* 4:75–82.

Walker, L. R., and del Moral, R. 2003. *Primary Succession and Ecosystem Rehabilitation.* Cambridge: Cambridge University Press.

Wardle, D. A., Walker, L. R., and Bardgett, R. D. 2004. Ecosystem properties and forest decline in contrasting long-term chronosequences. *Science* 305:509–513.

Watt, A. S. 1947. Pattern and process in the plant community. *Journal of Ecology* 35:1–22.

Weiher, E., and Keddy, P. A. 1995. The assembly of experimental wetland plant communities. *Oikos* 73:323–335.

Weiher, E., and Keddy, P.(eds.). 1999. *Ecological Assembly Rules: Perspectives, Advances, Retreats.* Cambridge: Cambridge University Press.

Whisenant, S. G. 1999. *Repairing Damaged Wildlands: A Process-orientated, Landscape-Scale Approach.* Cambridge: Cambridge University Press.

White, P. S., and Jentsch, A. 2001. The search for generality in studies of disturbance and ecosystem dynamics. *Progress in Botany* 62:399–450.

White, P. S., and Jentsch, A. 2004. Disturbance, succession, and community assembly in terrestrial plant communities. In: *Assembly Rules and Restoration Ecology: Bridging the Gap Between Theory and Practice.* V. M. Temperton, R. J. Hobbs, T. J. Nuttle, and S. Halle (eds.). Washington, D.C.: Island Press, pp. 341–366.

Wilson J. B. 1999. Guilds, functional types and ecological groups. *Oikos* 86:507–522.

Wilson, J. B., and Roxburgh, S. H. 1994. A demonstration of guild-based assembly rules for a plant community, and determination of intrinsic guilds. *Oikos* 69:267–276.

Young, T. P., Chase, J. M., and Huddleston, R. T. 2001. Community succession and assembly: Comparing, contrasting and combining paradigms in the context of ecological restoration. *Ecological Restoration* 19:5–18.

Zobel, M. 1997. The relative role of species pool in determining plant species richness: an alternative explanation of species coexistence? *Trends in Ecology and Evolution* 12:266–269.

8

Integrating Restoration and Succession

Richard J. Hobbs, Lawrence R. Walker, and Joe Walker

Key Points

1. Succession is a key ecological process that underpins much ecological restoration.
2. Restoration is a practical implementation of succession concepts that involves operational and field management control to decide where, what, how, and when to apply management actions to restore degraded ecosystems and landscapes.
3. The main components of restoration are planning (how to manipulate succession to a desired end-point); implementation (this can include applying disturbances in a predetermined way); and evaluation (deciding on the basis of expectations that the restoration is going in the right direction). All have links with succession and community assembly.
4. Practitioners of restoration are frequently not aware of the practical uses of succession concepts in planning and in setting targets. Communication from those working in succession to restoration practitioners and vice versa is needed to maximize benefits.

8.1 Introduction

In this book, we have explored the proposition that succession is a key ecological process that should underpin much ecological restoration. Ecological restoration is viewed as the manipulation of successional processes to meet realistic targets in restoring damaged landscapes. We have explored the linkages between ecological succession and the practice of restoration from a variety of different perspectives, including aboveground and belowground processes, and apply this understanding to restoration in particular biophysical settings and ecosystem types. The chapters address the question of what types of information are required from studies of ecological succession to aid decisions in restoration projects. In this chapter, we recognize that restoration activities must draw from a variety of disciplines within ecology as outlined in Table 8.1, but emphasize the strong linkage of restoration with succession. We group the types of information and actions needed for restoration under the set of questions first

Table 8.1 Fields of study that strongly influence restoration, the primary impact each has on restoration, and potential problems conducting restoration without addressing the contributions from each field.

Field	Impact	Problems if omitted
Succession	Predictions of temporal change and setting targets	Wrong trajectory identified for biodiversity outcomes, wrong species mixtures, arrested trajectory, improper manipulation of soil nutrients, goals too narrow
Assembly	Filters	Difficulty starting, mismatch of available propagules and the local environment
Landscape Ecology	Regional information exchange and extrapolation	Efforts limited to a single ecosystem or patch, poor estimates of role of surrounding biota, applying a "solution" to the wrong place
Disturbance Ecology	Initiation and boundaries	Multiple restarts, loss of biomass and soil organic matter from improper species choices, inadequate site stabilization
Climate Change	Shifting reference systems	Old communities no longer function as references or propagule sources, complications of novel communities
Historical Ecology	Legacies	Wrong targets set, belowground influences ignored
Environmental Ethics and Philosophy	Aesthetics, goals relating to naturalness and place	Societal values and expectations not achieved, visually unpleasant, ill-fitting or unachievable targets and lack of project support or maintenance

introduced in Chapter 1: why, where, what, how, and when? Explicitly, what are the perceived environmental or ecological issues leading to *why* restoration is being undertaken? What is the environmental domain and spatial extent *where* restoration will take place? How does the technical evaluation of *what* to expect help set end-points or targets following restoration? What evaluation of costs and benefits, operational control, resources, and knowledge is needed in order to decide *how* to implement the restoration program? Finally, *when* are physical manipulations of sites, vegetation establishment, ongoing reviews of progress, monitoring, and further interventions needed?

8.2 Why Is Restoration Needed?

It is often obvious why ecosystem, landscape, and mine site restoration is needed, given the visual evidence from most parts of the world of increasing impacts by humans and natural events on landscapes. These impacts have left large areas of land and water bodies polluted, reduced biodiversity and have dramatically decreased the use that can be made of our natural resources. Nevertheless, it is important to decide if and why restoration is required in particular contexts, and there are a variety of different biological and practical reasons why restoration activities are carried out. As discussed by Hobbs and Norton (1996), restoration is fundamentally conducted to improve or sustain ecosystem goods and services, which may include aesthetic and societal preferences. To achieve this broad goal, restoration activities may be required to reverse severe, localized disturbances such as tailings from mine sites, to reinstate productive capacity in degraded agricultural systems, to maintain or return conservation values in protected areas, or to reinstate broader landscape processes essential to the continuation of both rural and urban production and conservation enterprises. The scope for applying successional concepts to restoration activities was shown in the preceding chapters to vary greatly in each of these cases.

8.3 Where Is Restoration Done?

The "Where?" question is partly tied up with the "Why?" question. What sort of ecosystem or landscape are we dealing with and what successional stage is it in? Is the landscape old and does it need to be stabilized before it becomes irreversibly damaged? Is it a natural area where the management focus is on conservation, or is it a production landscape in which the primary focus is agriculture or forestry? Has the area been severely modified through mining or urbanization, or is the area simply in poor condition through mismanagement such as overgrazing? In other words, is the situation more akin to the relative infertility of the initial stage of primary succession or to secondary succession where a biological legacy remains? Studies of disturbance regimes and ecological succession can both help clarify answers to these fundamental questions.

Also involved in the "where?" question is the size of area being considered. For example, in Chapter 4 we see that the visual symptoms of salinity can be confined to a relatively small patch, but restoration is needed at a broader landscape scale, perhaps even across regions or state boundaries. Is it possible to consider single patches in isolation, or do we need to consider the broader landscape context in which the patches sit? For instance, if we are working with a degraded, semiarid woodland in Australia, is putting a fence around the woodland to reduce grazing pressure (as discussed in Chapter 3) going to be sufficient to restore the area, or do we need to deal with the broader landscape processes such as hydrological imbalance which might also be threatening the woodland (Cramer and Hobbs 2002)? In addition, dealing with particular problems in one location may lead to unintended impacts elsewhere; for instance, managing for salinity by increasing drainage in one area may lead to deleterious impacts downstream (Chapter 4). Similarly, using aggressive, fast-growing plant species as "pioneers" which rapidly colonize degraded areas may result in the species spreading into the surrounding landscape and becoming a problem weed (Parmenter *et al.* 1985, Sukopp and Starfinger 1999). Again, successional concepts can help plan appropriate restoration actions. Useful knowledge about local seed inputs and their dispersal mechanisms, species growth requirements and their interactions, and ecosystem impacts of focal species can be gleaned from successional studies, particularly if those studies have been conducted in similar habitats (see Chapter 2). Succession can also offer insights into likely outcomes on old landscapes (see Chapter 4), or help identify situations where unaided succession is the best restoration approach (see Chapter 6).

8.4 What Is Being Restored?

What to restore is one of the key questions in restoration projects, but it is often poorly addressed. Clear definition of the problem being tackled and the goal or end-point being considered is essential for the successful conduct and completion of restoration projects (see Chapter 1, Hobbs 1999). Successional studies can help decide what a restored ecosystem might "look like" but prediction can be inaccurate (see Chapters 2 and 7). Often the information comes from nearby "reference systems" or historical data to decide what the system was previously (Egan and Howell 2001; see Chapters 3, 4, and 7). These approaches generally consider the structural or species richness aspects of the system, whereas in many cases the functional aspects of the substrate are easier to restore

(Lockwood and Pimm 1999). Indeed, all the chapters in the book emphasize that structure and function go hand in hand, and that one cannot readily be restored without the other. However, one may not need all the species that were once in a particular location to successfully restore ecosystem function. Partial restoration of function, in turn, allows most species to return and persist.

The goals that are chosen also depend on a number of considerations, including the current state of the system and the degree of disturbance that has been experienced in the past and is currently being experienced. In some cases, the state of the system has been irrevocably changed and a return to the original may be impractical (see Chapters 3 and 7). In addition to restoring these intrinsic properties of the system, the levels of economic resources and societal commitment to the restoration are important determinants of what can realistically be achieved. These latter aspects are critical determinants, but are not fixed and can change with changing societal attitudes and political settings. A recent example of this is the situation in and around New Orleans (USA), where the extent of the damage from Hurricane Katrina in September 2005 was in some part attributable to the decline in coastal wetlands and barrier islands on the Louisiana coast. The problem of degrading coastal ecosystems had been recognized for some time, and ambitious but costly restoration projects had been proposed but rejected by the government prior to the hurricane. Post-Katrina, it is now thought that the government will be much more easily convinced that such projects should go ahead (Arnold 2006).

8.5 How Does Restoration Proceed?

The "How?" of restoration is intimately tied up with the "Why?" and "What?" questions. Deciding how to restore a degraded landscape involves selecting from the current and expanding set of management options available, given limitations set by the system to be restored, and the financial resources and expertise available to do it. Proximate considerations relate to the type and extent of damage being reversed. In some cases, the system can be left to regenerate on its own or simple inexpensive biotic manipulations may be all that is required. Alternatively, more expensive species introductions and plantings may be needed. In some cases, abiotic factors may need remediation first (Whisenant 2002, Chapters 5 and 6). Further considerations relate to the spatial scale of the restoration project. Clearly, small, patch-scale restorations will have very different methodologies and constraints from broad-scale landscape and regional restoration efforts. Nonetheless, even landscape restoration has to be conducted by treating individual landscape elements, and careful attention to spatial relations is essential. Guidance for making these restoration decisions comes from an understanding of successional dynamics at the site.

Broader considerations relate to the societal will and ability to undertake restoration and decisions about who should conduct the restoration. In many cases, it is either the company responsibility for the degradation or a government agency, but over the past two decades, more community groups and nongovernment organizations have become involved in restoration programs. Is anyone interested in doing anything, and is there sufficient time and money to conduct the restoration effectively? Is there sufficient political will and are there effective policy mechanisms in place to allow the restoration to proceed? Is there sufficient knowledge to undertake the restoration, and who holds that knowledge?

These questions involve consideration not only of scientific knowledge about the biophysical characteristics of the area, but also of local knowledge, "folk wisdom," and knowledge of indigenous people. Such nonquantitative knowledge may, and perhaps should, play an essential part in formulating the goals and determining the methodologies of the restoration project.

In many cases, the failure to use appropriate local and or scientific knowledge in restoration activities is due to a lack of communication between the theoreticians and the practitioners. Many practitioners may not realize such knowledge exists or have not even heard about historical land management practices or successional principles. Such knowledge, when couched in the context of helping to define and set restoration targets, can sometimes be productively assimilated into practical restoration activities.

8.6 When Should Restoration Occur?

This question relates to the timing of restoration efforts, both in terms of when the restoration should commence and when during the restoration further interventions are likely to be needed. The first simple answer to the question "When?" is "The sooner the better." It is much easier to maintain functioning systems than to repair damaged ones, and it is much more effective to repair damage early than to wait until things degrade further. This is particularly true in the face of mounting evidence that many systems are subject to threshold phenomena which involve sudden, nonlinear change from a less-degraded to a more-degraded state (Mayer and Rietkerk 2004, Suding *et al.* 2004; see Chapter 7). Such change is often difficult to reverse without costly intervention. Assessing the relative vulnerability of ecosystems to disturbance or land-use change can help prioritize which ecosystems need the most urgent attention, a job usually completed by land agencies.

The temporal aspects of restoration are numerous, and are intimately tied to successional processes. Temporal issues of restoration vary from the practicalities of ensuring that planting and seed sowing are carried out in the correct seasons to allow for germination and growth, to problems created by a run of dry or wet seasons and concerns over changing regional or even global environmental conditions at increasing temporal scales. This mix of timescales is also important when one considers the long-term implications of short-term actions. For instance, what long-term impacts will result from the introduction of a particular species to the system? A key element is to decide how and when short-term interventions will either set the system on the desired trajectory to the goal that has been set or will lead it off in another direction, perhaps with less desirable outcomes (see Chapters 3 and 6). This decision is also important in considering the way we measure the progress of the restoration. Many restoration projects require clear "completion criteria." In reality, it may take decades or centuries to reach the desired goal, but in the meantime, we have to make decisions on whether the course of development is appropriate and heading in the right direction.

An important aspect to consider is the effective monitoring of restoration projects to ensure progress toward a desired outcome is being achieved. Monitoring is also tied heavily into the process of adaptive management. Without effective monitoring, it will not be possible to assess whether any remedial or other management interventions are needed. While the need for effective monitoring is frequently recognized and advocated, the development of such monitoring regimes is often problematic. Monitoring has to be cost-effective, simple, targeted, and capable of actually detecting changes in the relevant parameters. It also has to be continued over a timeframe congruent with the progress of the restoration activity. As an example of what is required, Grant (2006) has recently presented a successional model for mine rehabilitation, using a state and transition approach that considers the monitoring requirements at each stage of the restoration process.

8.7 Lessons from Succession

What can succession, the science of species and substrate change, provide that will aid in making the range of decisions outlined above? The case that information from succession is fundamental to all restoration is compelling but we recognize that this view has not been prominent among practitioners of restoration. Perhaps succession would be more acceptable as a tool for restoration if it were viewed as a unifying concept, linking disturbance, assembly, and landscape ecology to restoration activities. Further, we believe that restoration is not currently getting what it needs from those studying succession because of cultural and conceptual differences between the two areas (Table 8.2). However, we are hopeful that many of these differences can be overcome with proper communication because we think the urgent process of restoration needs the insights from ecology, and succession in particular. Similarly, our understanding of successional dynamics can be vastly improved by incorporating the experiential lessons from restoration.

We address three areas where succession can aid restoration: planning, species trajectories, and ecosystem functions. Each of these three areas has close links to the 3 R's of restoration as depicted in Fig. 8.1 ("reading" = planning; "writing" = monitoring and manipulating species trajectories;

Table 8.2 Questions and answers about the linkage between restoration and succession. Our preferred answers are in bold, followed by possible explanations.

Question 1: Does restoration (R) need succession (S)?

NO: a: R has its own methodology ("bypasses" succession)
       b: R needs are different
       c: S not a useful model when dealing with noxious invasives

YES: a. R is a manipulation of S so its needs are similar
       b. R is the engineering arm of S
       c. S explains the key principles needed for R
       d. S can fine tune R
       e. S can reduce errors for reaching R targets

Question 2: Does R currently draw from knowledge of S to satisfy needs?

NO: a. Cultural differences among R and S personnel
       b. Concepts about endpoints differ
       c. Communication between R and S poor
       d. Perceived lack of useful information from S
       e. Most restoration personnel have not heard of succession
       f. Technical information transfer from S to R is poor

YES: a. Climax concept widespread, at least implicitly
       b. Goals set within biologically feasible limits
       c. Species dynamics and ecosystem functions addressed

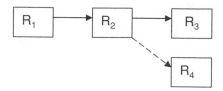

Figure 8.1 The three "R's" of restoration—sequential restoration actions over time. R₁: "Reading" (assessing) the situation. Is restoration needed? Can it be done? What can it accomplish? What are its goals? What minimum amount of information is needed for it to begin? What is the nature of that information? R₂: "Writing" (doing) restoration via monitoring, maintenance, manipulating as needed to revise trajectories (R3 versus R4) or remove bottlenecks, and responding to changing disturbance regimes and environmental changes. R₃,₄: "Arithmetic" (evaluating) of restoration by measuring success at achieving goals and by modeling results to provide generalities and lessons for future restoration activities.

"arithmetic" = evaluating restoration success, e.g., by attainment of certain ecosystem functions).

8.7.1 Planning

An old adage is that anything is possible if you throw enough money and resources at it. In the case of restoration, this is only true if the activities do, in fact, point restoration trajectories in the correct direction. Because money is rarely available in an endless supply, decisions are needed on how to spend the meager dollars most efficiently to achieve the aim. Restoration targets vary in purpose from very narrow (one endangered species), to constrained (a particular species), to narrow (a focal community), to moderate (a certain level of biodiversity), to wide (certain growth forms, dominant species or ecosystem functions). Successional studies can help in the restoration process (Fig. 8.1), to define the broad goals or specific end-points and to choose which path to take to reach the goals. In the planning phase, there are many opportunities for inputs from succession. The initial phase (R₁) is a period of assessment about the utility of restoration and the critical phase of defining ecologically sound restoration goals. Initial site amelioration is often needed (e.g., to stabilize or adjust nutrients in the substrate) and experience from successional studies (e.g., about likely guilds or functional groups to introduce) can provide guidance for restoration planning. Long-term goals can be refined to include both abiotic and biotic structures and system states that are most likely to be achievable. As a planning mechanism, there are many computer-based tools that can be utilized to display a range of scenarios to assist both the decision making and implementation phases of restoration. Such applications should aim to minimize restoration effort and initiate, design, and suggest short-term interventions that manipulate the long-term successional process.

There is often a range of potential restoration goals in any given situation and the decision about which goal to aim for can be difficult. Decisions will be based on an array of factors such as available resources, ecosystem type, and societal attitudes toward the sustainability of particular ecosystems or the goods and services provided. Information about long-term succession can improve the certainty of particular management actions and reduce the possibility of error or undesirable variants. For instance, restoration planning needs to include multiple growth forms and functional groups to promote spatial mo-saics of community types and use local mature vegetation as a guide but not as an exact model for restoration (see Chapter 2). Involvement of soil biologists in restoration planning can address impacts of aboveground manipulations on belowground processes (see Chapter 3). There is also a need to recognize the age of landscapes when stabilizing agricultural systems (see Chapter 4) and realize that disturbance intensity strongly impacts recovery pathways and often leads to multiple restoration scenarios, especially in cultural landscapes such as in Europe (see Chapter 5). Further, one must always determine the minimal restoration effort needed to achieve desirable results, even if that means doing nothing at all (other than removing the source of disturbance or degradation) (see Chapter 6). Many other topics in the preceding chapters address additional issues involved in restoration planning (e.g., how to incorporate or exclude invasive species, thicket-forming competitors, appropriate fertilization levels, strategic grazing or exclusion of grazers, rewetting wetlands, defining achiev-able end-points, or gaps in knowledge). These various perspectives suggest the value of considering all applicable inputs from successional knowledge before beginning any restoration activities.

8.7.2 Species Trajectories

Once restoration is underway (Fig. 8.1, R_2), there is ample opportunity for inputs from succession to modify trajectories and plant community composition and to verify if ecosystem development is conforming to expectations. Much work in succession has focused on the dynamics of species replacements and resulting community properties such as assembly, composition, and structure. Understanding succession can clarify the most likely impacts of nurse plants for facilitation or species combinations where competition will impede restoration. For example, shrubs in the succulent karoo habitat of South Africa were more likely to inhibit seedlings under dry, nutrient-rich conditions than under either wet or dry, nutrient-poor conditions (Riginos et al. 2005). In many restoration sites, excessive fertilization leads to plant monocultures and reduction in species diversity (see Chapters 2 and 6).

An understanding of succession can aid restoration actions by defining a framework of possible trajectories within which restoration can occur (Fig. 8.2; see Chapter 6). If restoration exceeds those boundaries, it will likely fail (unless other endpoints are deemed acceptable). Returning to the most direct path to the target will reduce time and costs. Knowledge about succession is also useful in selecting vulnerable points in the developmental sequence when strategic restorative manipulations can be most efficient and effective. These vulnera-ble points are often when there is a shift in dominant species or life forms (see Chapter 7) and the community is most susceptible. Alternatively, succes-sional insights may help the practitioner of restoration overcome bottlenecks by identifying their causes (e.g., impermeable hardpan and water logging) and suggesting solutions (e.g., by modifying the water cycle to harvest surplus wa-ter and improve the organic content of surface soil; see Chapters 4 and 5). In addition, information about successional changes can suggest how to increase the rate of change, e.g., by adding critical soil nutrients or appropriate mycor-rhizae (see Chapter 3) or how to enlarge the final target (Walker and del Moral 2003; see Chapter 6).

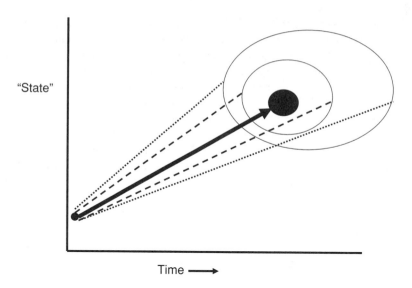

"State"

Time ⟶

Figure 8.2 Classical restoration trajectory (solid line) and increasingly wider targets (dashed and dotted lines) reflecting unpredictability or variability of endpoints and multiple potential outcomes for restoration. Options can expand or contract through the restoration process.

Precise restoration of original species assemblages and/or ecosystem functions is almost always unrealistic because of the dynamic nature of disturbance and changing reference systems. If we think in terms of restoration trajectories (Fig. 8.2), the general aim in restoration is to move the system from its current (degraded) state to a stable target state. Diagrams usually depict this with a single arrow joining the current and target states. However, it has to be acknowledged that there are numerous factors that affect the predicted direction and cohesiveness of the trajectory. These include factors that are beyond the control of project managers, such as prolonged drought, and practical issues that are part and parcel of fieldwork, including delays in supply or poor quality materials. Options may contract due to unexpected loss of species (e.g., from surrounding habitat destruction, species extinction) or due to an inability of the soil and belowground biotic systems to respond (e.g., crossing thresholds for recovery for soil structure). Options may also increase if new species (e.g., weeds) colonize, land use changes, or societal views change to be more accepting and even demanding of novel rather than original systems (Hobbs *et al.* 2006). Restoration trajectories will also undoubtedly be impacted by climate change but in an unknown way (Harris *et al.* 2006). Hence trajectories are likely to be more dynamic than usually depicted, and options may expand or contract along the way. Adaptive management strategies are the best way to deal with divergence from anticipated end-points. Although in many cases restoration trajectories are less predictable than we would like, we suggest that knowledge from succession theory can reduce the unpredictability. Restoration needs to move toward an evidence-based paradigm and as pointed out earlier, evidence can come from many sources.

The final stage of a restoration that aims for complete species replacement is rarely achieved (see Chapters 2, 5, and 6). First, broad scale disturbances can remove species pools and the ecosystem may have changed to a new state that does not provide the resource base for the original species (see Chapter 4). Second, use of ecosystems for goods and services can change in the future and so yesterday's goals may no longer seem reasonable. Third, in some cases restoration targets are years away. Finally, restoration targets are likely to change as novel species combinations appear, invasive organisms dominate, native plants become rare or extinct, or climate changes. All of these challenges suggest that multiple possible targets with broadly defined parameters are most likely to succeed (Fig. 8.2). These targets may include a mosaic of habitats or functional groups of species within a locale.

8.7.3 Ecosystem Functions

The goal of ecological restoration is to establish a self-sufficient ecosystem that requires minimal or no continuing human inputs in order to provide a continuing supply of goods and services. The more visible structural components (e.g., vegetation height, cover or biomass, species richness and distribution) are often emphasized at the cost of ensuring the proper functioning of the abiotic components (e.g., nutrient dynamics, productivity, the water cycle, soil structure, decomposition) . The two aspects are interdependent because manipulations of one inevitably result in change in the other.

Evaluation of the success at achieving restoration goals (Fig. 8.1; $R_{3,4}$) is frequently carried out by monitoring easily measured parameters of standardized ecosystem functions. Ecosystem dynamics are complex, but by assuming some processes dominate in particular locations it is possible to develop a set of indicators to assess the direction of change following management actions (Walker and Reuter 1996; see Chapter 5). In some cases the focus may be on terrestrial measures, while in others impacts of land-use or changes to land-use are better measured in streams or other water bodies. Some parameters of soil fertility, for example, are easily measured and provide a critical gauge of soil development and nutrient dynamics (see Chapter 3). Measures of subsets of ecosystem functions across successional seres or stages can provide values to help establish benchmarks for restoration management and evaluation. These measures can also be used as input variables for process-based models (see Chapter 5).

Long-term field studies of successional changes can also help place a timeline for recovery and identify critical points where further intervention is needed. For example, recovery of tropical forests on abandoned, paved roads in Puerto Rico was characterized by the sequential recovery of soil pH, soil organic matter, litter mass, soil moisture, soil nitrogen, and, finally, species richness (Heyne 2000). Species composition did not return to control levels in the 80-year chronosequence.

8.8 Conclusions

We have focused on why lessons learned about species and ecosystem change from studies of plant succession are central to the practice of restoration. However, restoration has much to assist the study of succession (see Chapter 1). Bradshaw (1987) and others have observed that restoration activities offer an unprecedented test of a host of ecological ideas by providing experimental manipulations, often on a scale only dreamt of by experimental ecologists. Careful use of restoration activities, past and present, by ecologists can help inform continuing debates over key processes and drivers of succession and other important ecological phenomena.

Successful integration of theory and practice is an ongoing challenge that faces many hurdles, as indicated previously. Succession has been, and still mostly remains, a conceptual construct that aids in the understanding of how systems change over time. Restoration is primarily a practical activity that seeks to achieve outcomes in the most effective and efficient way possible. Many restoration practitioners remain unconvinced that they need a more general body of theory to guide their work because much of what they do is driven by the local idiosyncrasies of the site and system they are working on. And yet our ability to build on local activities and transfer knowledge from one project to another depends on our ability to set each individual activity in a broader context—and this demands a set of concepts which can act as reference points against which particular activities can be compared.

The field of succession has much to offer in this direction, as we have tried to elaborate in this book. However, the successful transfer of successional knowledge into the practical realm of restoration demands a willingness of scientists and practitioners to engage in effective dialogue: that involves both talking and listening on both sides. Scientists have to get their own house in order and decide which concepts are useful, which overlap, and which are redundant. They then need to translate these concepts into practical tools that can be readily picked up and used by practitioners. The history of research in succession spans well over a century and we are still debating concepts and models. The history of restoration is much shorter, but the need for effective restoration is rapidly increasing. While debate on concepts and models is necessary and important, the time is ripe for the effective transfer of these concepts to the practicalities of restoration so that we can ensure that restoration efforts are as effective and efficient as they need to be.

Acknowledgements: Comments by Tadashi Fukami, Truman Young, and Joy Zedler greatly improved this chapter. Lawrence Walker acknowledges sabbatical support from the University of Nevada Las Vegas and from Landcare Research, Lincoln, New Zealand.

References

Arnold, G., ed. 2006. *After the Storm: Restoring America's Gulf Coast Wetlands*. Washington, D.C.: Environmental Law Institute, pp. 62.

Bradshaw, A. D. 1987. Restoration: An acid test for ecology. In: *Restoration Ecology: A Synthetic Approach to Ecological Research*. W. R. Jordan, M. E. Gilpin, and J. D. Aber (eds.). Cambridge: Cambridge University Press, pp. 23–30.

Cramer, V. A., and Hobbs, R. J. 2002. Ecological consequences of altered hydrological regimes in fragmented ecosystems in southern Australia: Impacts and possible management responses. *Austral Ecology* 27:546–564.

Egan, D., and Howell, E. A. (eds.). 2001. *The Historical Ecology Handbook: A Restorationist's Guide to Reference Ecosystems*. Washington, D.C.: Island Press.

Grant, C. D. 2006. State-and-transition successional model for bauxite mining rehabilitation in the jarrah forest of Western Australia. *Restoration Ecology* 14:28–37.

Harris, J. A., Hobbs, R. J., Higgs, E., and Aronson, J. 2006. Ecological restoration and global climate change. *Restoration Ecology* 14:170–176.

Heyne, C. 2000. Soil and vegetation recovery on abandoned paved roads in a humid tropical rainforest, Puerto Rico. M.S. Thesis, University of Nevada Las Vegas.

Hobbs, R. J. 1999. Restoration of disturbed ecosystems. In: *Ecosystems of Disturbed Ground. Ecosytems of the World 16*. L. Walker (ed.). Amsterdam: Elsevier, pp. 673–687.

Hobbs, R. J., Arico, S., Aronson, J., Baron, J. S., Bridgewater, P., Cramer, V. A., Epstein, P. R., Ewel, J. J., Klink, C. A., Lugo, A. E., Norton, D., Ojima, D., Richardson, D. M., Sanderson, E. W., Valladares, F., Vilà, M., Zamora, R., and Zobel, M. 2006. Novel ecosystems: Theoretical and management aspects of the new ecological world order. *Global Ecology and Biogeography* 15:1–7.

Hobbs, R. J., and Norton, D. A. 1996. Towards a conceptual framework for restoration ecology. *Restoration Ecology* 4:93–110.

Lockwood, J. L., and Pimm, S. L. 1999. When does restoration succeed? In: *Ecological Assembly Rules: Perspectives, Advances, Retreats*. E. Weiher, and P. Keddy (eds.). Cambridge: Cambridge University Press, pp. 363–392.

Mayer, A. L., and Rietkerk, M. 2004. The dynamic regime concept for ecosystem management and restoration. *BioScience* 54:1013–1020.

Parmenter, R. P., MacMahon, J. A., Waaland, M. E., Stuebe, M. M., Landres, P., and Crisafulli, C. 1985. Reclamation of surface coal mines in western Wyoming for wildlife habitat: A preliminary analysis. *Reclamation and Revegetation Research* 4:98–115.

Riginos, C., Milton, S. J., and Wiegand, T. 2005. Context-dependent interactions between adult shrubs and seedlings in a semi-arid shrubland. *Journal of Vegetation Science* 16:331–340.

Suding, K. N., Gross, K. L., and Houseman, G. R. 2004. Alternative states and positive feedbacks in restoration ecology. *Trends in Ecology and Evolution* 19:46–53.

Sukopp, H., and Starfinger, U. 1999. Disturbance in urban ecosystems. In: *Ecosystems of Disturbed Ground. Ecosystems of the World 16*. L.R. Walker (ed.). Amsterdam: Elsevier, pp. 397–412.

Walker, J., and Reuter, D. J. 1996. Indicators of catchment health: A technical perspective. Collingwood, Australia: CSIRO Publishing.

Walker, L. R., and del Moral, R. (eds.). 2003. *Primary Succession and Ecosystem Rehabilitation*. Cambridge: Cambridge University Press.

Whisenant, S. G. 2002. Terrestrial systems. In: *Handbook of Ecological Restoration* Volume 1: *Principles of Restoration*. M. R. Perrow, and A. J. Davy (eds.). Cambridge: Cambridge University Press, pp. 83–105.

Glossary

abiotic. Pertaining to nonbiological factors such as soil, water, wind, temperature

aeolian deposits. Wind-borne, soil-forming particles such as sand or dust

alien species. A species from another biome or continent; a nonnative organism (see exotic species)

alternative stable states. Different configurations that a particular ecosystem can occur in depending on the interplay of past conditions and disturbance events

anthropogenic. Caused by humans

assembly rules. Predictions concerning mechanisms of community organization

basophilous. Thriving in alkaline habitats

biodiversity. Number and distribution of species

biogeography. Study of the distribution of organisms

biome. A geographical region with similar vegetation and climate

bioremediation. Reclamation based on the use of plants to reduce toxicity

biotic. Pertaining to biological factors

bog. A wetland with low pH, usually saturated soil and dominance by mosses

C. Carbon

calciphilous. Thriving in environments rich in calcium salts

carbon sequestration. The uptake and storage of carbon as through photosynthesis

carr. A wetland with deciduous trees such as alder or willow

catchment. The area of land drained by a river system

chronosequence. A series of communities arrayed on the landscape and presumed to represent a successional sequence (a space-for-time substitution)

climax vegetation. Vegetation that has reached a stable state (dynamic equilibrium); the optimum expression of vegetation for the climate and soils of a region

competition. The negative influence of one species on another due to sharing of limited resources

denitrification. The bacterial reduction of nitrate to nitrogen under anaerobic conditions

direct regeneration. Regrowth of previous vegetation following disturbance with no intervening successional stages

discharge. The water that has moved into groundwater and comes out at low points of the landscape forming wet areas (springs); often contains pollutants such as salt

disturbance. A relatively discrete event in time and space that alters habitat structure and often involves a loss of biomass or soil

disturbance regime. The composite influence of all disturbances at a particular site

ecosystem function. Processes that define the workings of an ecosystem such as carbon sequestration, nutrient dynamics, or water flow

ecosystem service. Usefulness of an ecosystem to society, such as providing clean water

ecosystem structure. Physical aspects of an ecosystem including biomass, plant cover, species density

ecohydrology. That aspect of hydrology focusing on ecological aspects such as transpiration and energy balance

ecotone. The transition zone between two communities

eutrophication. The process by which an aquatic system becomes more fertile; usually a negative result ensues

evapotranspiration. Total water loss per unit area from both evaporation from soil and water surfaces and transpiration from plant surfaces

exergy. Work (usable energy) potentially extractable from physical systems

exotic species. Species not native to the location; often a weed (see alien species)

evenness. Relative abundance of species; with richness, a component of species diversity

facilitation. The positive influence of one species on another in a successional or restoration context

fen. An oligotrophic, acidic habitat dominated by herbaceous species, not mosses; frequently saturated

functional group. Species that share physiological, morphological, or behavioral traits

functional redundancy. When two or more species perform similar roles in an ecosystem

gap dynamics. Replacement of individuals in small disturbances within a largely intact matrix

glacial foreland. The terrain exposed by a receding (melting) glacier

guild. Suite of species with similar functional properties

guild proportionality. Within a particular community type, the proportions of species of different guilds are almost constant across sites in different developmental stages

habitat heterogeneity. Diversity of habitats within an ecosystem

heathland. A sandy, acidic, infertile habitat dominated by small-leaved shrubs

herbivore. An organism that eats plant parts

hysteresis. The impact of historical effects on the current response of an ecosystem to disturbance

inhibition. Any mechanism by which one species reduces the success of another in a successional or restoration context

LAI (leaf area index). The area of green leaf per unit area of ground surface

landscape ecology. The study of interactions of physical and biological phenomena across large spatial scales and multiple ecosystems

laterite. A tropical soil with high clay content and pronounced leaching resulting in accumulation of iron and aluminum layers at depth

life history characteristics. The species-specific patterns of arrival, growth, and longevity

luxury consumption. Uptake of resources by a plant beyond what it currently needs

macroclimate. The climate of a large region

mesocosm. Experimental system that closely resembles real-life conditions

mesotrophic. Moderately productive lakes (between eutrophic and oligotrophic)

microbe. A microscopic organism

microclimate. Small-scale climate, such as around a seedling

microsite. Small-scale habitat (see safe-site)

microtopography. Small-scale physical features of the land such as furrows or ridges

mineralization. The breakdown of organic compounds into inorganic chemicals

mine tailings. The wastes remaining after extraction of minerals or fossil fuels

mire. A peat-forming ecosystem such as a fen or a bog

moorland. An infertile, peaty habitat dominated by small-leaved shrubs, ferns, and mosses

mor. A forest soil type with discrete humus layer produced under cool moist conditions

mutualism. A biotic interaction among different species that is beneficial to both

mycorrhizae. Fungi that form mutualistic interactions with higher plants

N. Nitrogen

nematode. Generally microscopic, unsegmented worms often parasitic on plant roots

net primary productivity (NPP). The sum of all plant biomass generated in a given time and place

nurse plant. An established individual that alters its immediate surroundings in ways that favor establishment of another plant

oligotrophic. An unproductive lake

P. Phosphorus

pathogen. A disease-producing organism

peat. Slightly decomposed organic material accumulated under conditions of excess moisture

pedogenesis. The formation of soils

phenotypic plasticity. Nongenetic variation in organisms in response to environmental factors

phytomass. Plant biomass

podzol. A soil profile with extensive leaching of minerals to the lower B horizon

priority effects. The consequences of arrival order that condition subsequent compositional changes

propagule. Any reproductive unit that is adapted to dispersal

reallocation. The conscious transformation of a landscape to a condition or use distinct from its original one

recalcitrant. Not easily decomposed (as in recalcitrant leaf litter)

recharge. The amount of water that goes below the root zone and into groundwater

reclamation. The conversion of wasteland to some productive use by conscious intervention

rehabilitation. Any manipulation of a sere to enhance its rate or to deflect its trajectory toward a specified goal

releve. A subjectively chosen plot to sample relatively homogeneous plant communities

rescue effect. Immigration of individuals likely to reproduce, thereby saving a local population from extinction

resilience. Capacity to recover following disturbance

restoration. Returning the land to its exact (sensu stricto) or approximate (sensu lato) biological status

retrogressive. See trajectory

riparian. Pertaining to growth along a river corridor

ruderal. A weedy plant that colonizes recent disturbances

safe site. A microsite where seeds have an enhanced chance to lodge, germinate and establish

seral stage. One successional stage (see sere)

sere. A successional sequence, including all (seral) stages

stability. A community characteristic expressing lack of change or resistance to disturbance

stochasticity. Unpredictability

succession. Species change over time

primary succession. Species change following removal of most plants and soil

secondary succession. Species change following a disturbance that leaves soil layers relatively intact

spontaneous succession. Unmanipulated succession, as in abandonment (in lieu of purposeful restoration activities)

threshold. A point at which a small change in a driving variable can cause a large, potentially irreversible change in the state of an ecosystem

trajectory. The temporal path of vegetation from its initiation to stability

> **arrested.** The development of a sere is delayed in response to factors such as dominance by one species

> **convergent.** A sere develops increasing similarity to a local mature community or two seres become increasingly similar

> **divergent.** Two communities become increasing dissimilar over time

> **progressive.** An increase in stature, biomass or biodiversity over time

> **retrogressive.** A reduction in stature, biomass or biodiversity of an ecosystem due to erosion, leaching of nutrients or other disturbance

Index